The ESSE[illegible]® of

Biology I

Staff of Research and Education Association
Dr. M. Fogiel, Director

This book covers the usual course outline of Biology I. For more advanced topics, see *"THE ESSENTIALS OF BIOLOGY II."*

Research & Education Association
61 Ethel Road West
Piscataway, New Jersey 08854

THE ESSENTIALS®
OF BIOLOGY I

Year 2002 Printing

Printed in the United States of America

Library of Congress Control Number 00-130598

International Standard Book Number 0-87891-573-7

IV-2

WHAT "THE ESSENTIALS" WILL DO FOR YOU

This book is a review and study guide. It is comprehensive and it is concise.

It helps in preparing for exams and in doing homework, and remains a handy reference source at all times.

It condenses the vast amount of detail characteristic of the subject matter and summarizes the **essentials** of the field.

It will thus save hours of study and preparation time.

The book provides quick access to the important facts, principles, theorems, concepts, and equations in the field.

Materials needed for exams can be reviewed in summary form—eliminating the need to read and re-read many pages of textbook and class notes. The summaries will even tend to bring detail to mind that had been previously read or noted.

This "ESSENTIALS" book has been prepared by an expert in the field, and has been carefully reviewed to ensure accuracy and maximum usefulness.

Dr. Max Fogiel
Program Director

CONTENTS

CHAPTER 1

The Chemical and Molecular Basis of Life

1.1 The Elements

Element – An element is a substance which cannot be decomposed into simpler or less complex substances by ordinary chemical means.

Compound – Compounds are a combination of elements present in definite proportions by weight. These are substances which can be decomposed by chemical means.

Mixtures – Mixtures contain two or more substances, each of which retains its original properties and can be separated from the others by relatively simple means. They do not have a definite composition.

Atoms – Each element is made up of one kind of atom. An atom is the smallest part of an element which can combine with other elements. Each atom consists of:

A) atomic nucleus – small, dense center of an atom.

B) proton – positively charged particle of the nucleus.

C) neutron – electrically neutral particle of the nucleus.

D) electron – negatively charged particle which orbits the nucleus.

In normal, neutral atoms, the number of electrons is equal to the number of protons.

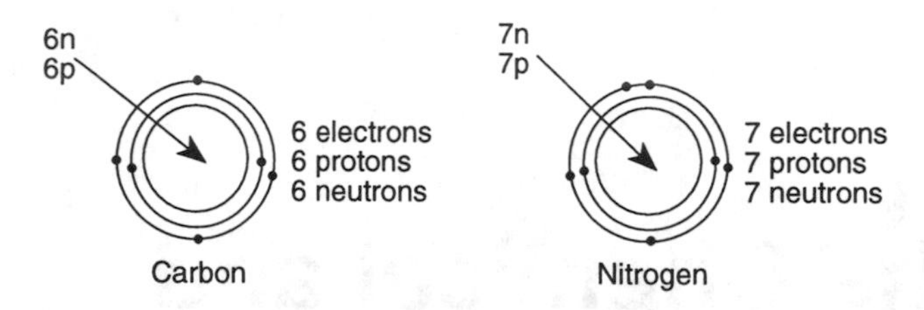

Figure 1.1 Atomic structure of carbon and nitrogen

Atomic Mass – The total number of protons and neutrons in a nucleus is the atomic mass. This number approximates the total mass of the nucleus.

Atomic Number – The atomic number is equal to the number of protons in the nucleus of an element.

Isotope – Atoms of the same element that have a different number of neutrons are known as isotopes. All isotopes of the same element have essentially the same chemical properties but their physical properties may be affected.

Ions – Atoms or groups of atoms which have lost or gained electrons are called ions. One of the ions formed is always electrically positive and the other electrically negative.

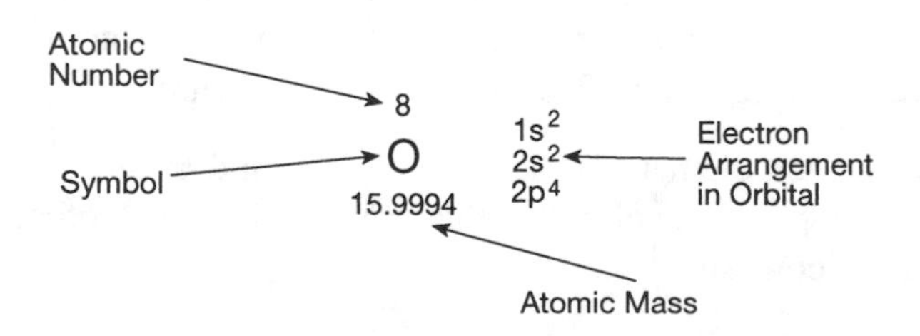

Figure 1.2 Representation of oxygen from Periodic Table of the Elements

1.2 Chemical Bonds

Ionic Bond – This involves the complete transfer of an electron from one atom to another. Ionic bonds form between strong electron donors and strong electron acceptors. Ionic compounds are more stable than the individual elements.

Covalent Bond – This involves the sharing of pairs of electrons between atoms. Covalent bonds may be single, double, or triple.

Polar Covalent Bond – A polar covalent bond is a bond in which the charge is distributed asymmetrically within the bond.

Non-Polar Covalent Bond – A non-polar covalent bond is a bond where the electrons are distributed equally between two atoms.

Hydrogen Bond – A hydrogen bond is the attraction of a hydrogen atom, already covalently bonded to one electronegative atom, to a second electronegative atom of the same molecule or adjacent molecule. Usually these bonds are found in compounds that have strong electronegative atoms such as oxygen, fluorine, or nitrogen.

Van der Waals Forces – Van der Waals forces are weak linkages which occur between electrically neutral molecules or parts of molecules which are very close to each other.

Hydrophobic Interactions – Hydrophobic interactions occur between groups that are insoluble in water. These groups, which are non-polar, tend to clump together in the presence of water.

1.3 Acids and Bases

Acid – An acid is a compound which dissociates in water and yields hydrogen ions [H^+]. It is referred to as a proton donor.

Base – A base is a compound which dissociates in water and yields hydroxyl ions [OH^-]. Bases are proton acceptors.

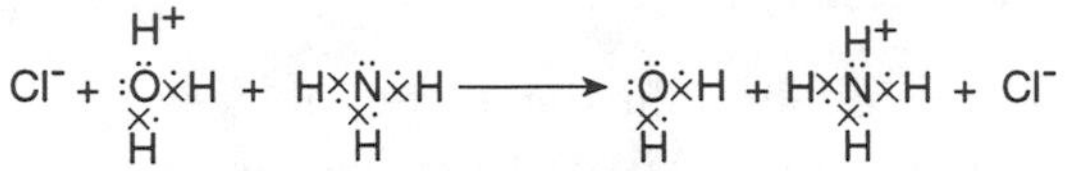

$$
\begin{array}{ccccccccccc}
 & & H^+ & & & & & & H^+ & & \\
Cl^- & + & :\ddot{O}\times H & + & H\times\ddot{N}\times H & \longrightarrow & :\ddot{O}\times H & + & H\times\ddot{N}\times H & + & Cl^- \\
 & & \times\!\cdot & & \times\!\cdot & & \times\!\cdot & & \times\!\cdot & & \\
 & & H & & H & & H & & H & &
\end{array}
$$

Figure 1.3 Reaction between hydrochloric acid (proton donor) and ammonia (proton acceptor)

pH – The degree of acidity or alkalinity is measured by pH.

$$pH = \frac{1}{\log [H^+]} = -\log [H^+]$$

$$pH = 7 \rightarrow \text{neutral}$$
$$pH < 7 \rightarrow \text{acidic}$$
$$pH > 7 \rightarrow \text{basic}$$

1.4 Chemical Changes

Chemical Reaction – A chemical reaction refers to any process in which at least one bond is either broken or formed. The outcome of a chemical reaction is a rearrangement of atoms and bonding patterns.

Laws of Thermodynamics

A) First Law of Thermodynamics (conservation of energy) – In any process, the sum of all energy changes must be zero.

B) Second Law of Thermodynamics – Any system tends toward a state of greater entropy meaning randomness or disorder.

C) The Third Law of Thermodynamics – A perfect crystal which is a completely ordered system, at absolute zero (0 Kelvin) would have perfect order, and therefore its entropy would be zero.

Stability of chemical system depends on:

A) enthalpy – total energy content of a system.

B) entropy – energy distribution.

Exergonic Reaction – Exergonic reactions release free energy; all spontaneous reactions are exergonic.

Endergonic Reaction – Endergonic reactions require the addition of free energy from an external source.

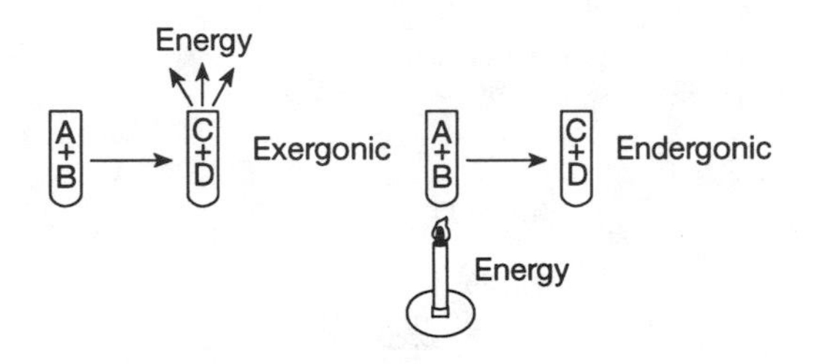

Figure 1.4 Exergonic and Endergonic Reactions

1.5 Organic Chemistry

A) Hydrocarbons – The simplest organic molecules are the hydrocarbons. These compounds are composed solely of carbon and hydrogen. They can exist as chains (e.g., butane) or rings (e.g., benzene).

```
                                          H
                                          |
    H   H   H   H                         C
    |   |   |   |                     //     \
H - C - C - C - C - H            H - C         C - H
    |   |   |   |                    |         ||
    H   H   H   H                H - C         C - H
        butane                         \\     /
                                          C
                                          |      benzene
                                          H
```

Figure 1.5

B) Lipids – Lipids are organic compounds that dissolve poorly, if at all, in water (hydrophobic). All lipids (fats and oils) are composed of carbon, hydrogen, and oxygen where the ratio of hydrogen atoms to oxygen atoms is greater than 2:1. A lipid molecule is composed of 1 glycerol and 2 fatty acids.

Phospholipid – A phospholipid is a variety of a substituted lipid which contains a phosphate group.

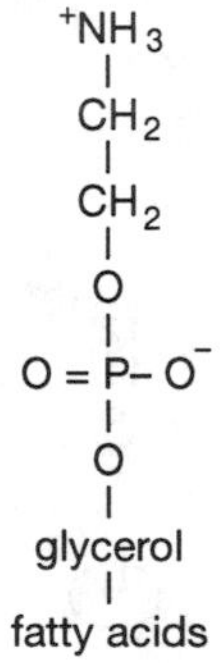

$^{+}NH_3$
|
CH_2
|
CH_2
|
O
|
$O = P - O^-$
|
O
|
glycerol
|
fatty acids

Figure 1.6 (A) An example of the addition which makes a lipid a phospholipid

C) Steroids – Steroids are complex molecules which contain carbon atoms arranged in four interlocking rings. Some steroids of biological importance are vitamin D, bile salts, and cholesterol.

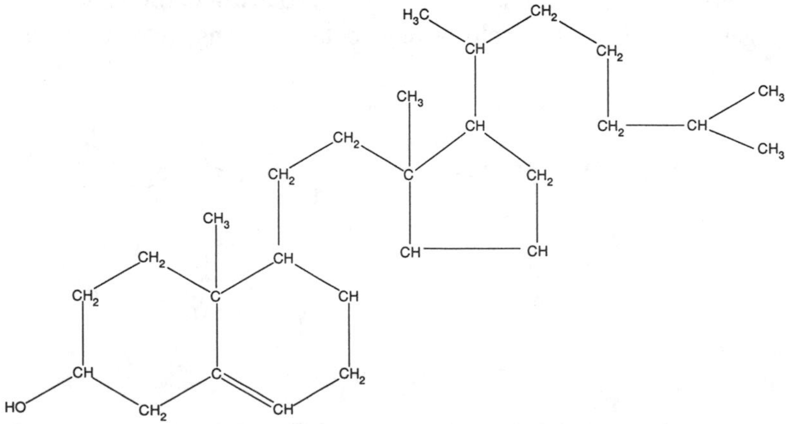

H_3C CH_2 CH CH_2 CH_3 CH CH_2 CH CH_3 CH_2 CH_3 CH_2 C CH_2 CH_3 CH CH CH_2 CH CH_2 C CH CH_2 CH C CH_2 HO CH_2 CH

Figure 1.6 (B) Structural formula of cholesterol

D) Carbohydrates – Carbohydrates are compounds composed of carbon, hydrogen, and oxygen, with the general molecular formula CH_2O. The principal carbohydrates include a variety of sugars.

1. monosaccharides – a simple sugar or a carbohydrate which cannot be broken down into a simpler sugar. Its molecular formula is $(CH_2O)_n$ and the most common is glucose $(C_6H_{12}O_6)$.

OH — OH

Figure 1.7 Glucose

2. disaccharide – a double sugar, or a combination of two simple sugar molecules. Sucrose is a familiar disaccharide as are maltose and lactose.

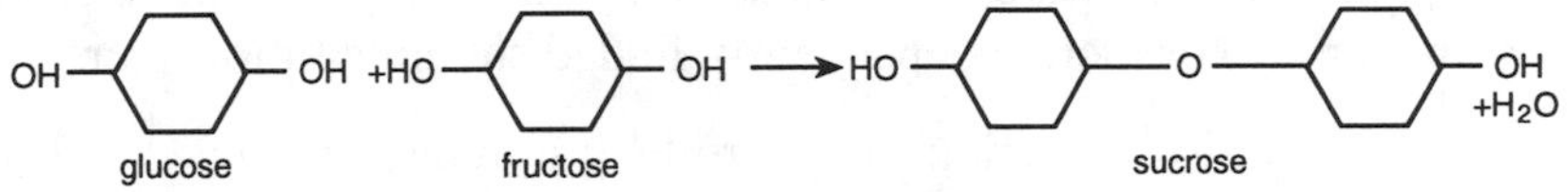

OH — OH +HO — OH → HO — O — OH $+H_2O$

glucose fructose sucrose

Figure 1.8 Double sugar formation by dehydration synthesis

3. polysaccharide – a polysaccharide is a complex compound composed of a large number of glucose units. Examples of polysaccharides are starch, cellulose, and glycogen.

E) Proteins – All proteins are composed of carbon, hydrogen, oxygen, nitrogen, and sometimes phosphorus and sulfur. Approximately 50% of the dry weight of living matter is protein.

Amino Acids – The twenty amino acids are the building blocks of proteins.

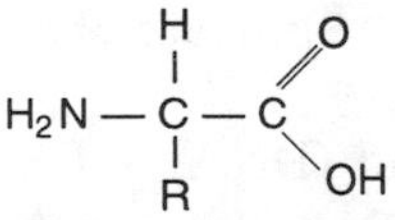

H_2N — C(H)(R) — C(=O)OH

Figure 1.9 An amino acid with R representing its distinctive side chain

Polypeptides – Amino acids are assembled into polypeptides by means of peptide bonds. This is formed by a condensation reaction between the COOH groups and the NH_2 groups.

Primary structure – The primary structure of protein molecules is the number of polypeptide chains and the number, type, and sequence of amino acids in each.

Secondary structure – The secondary structure of protein molecules is characterized by the same bond angles repeated in successive amino acids which gives the linear molecule a recurrent structural pattern.

Tertiary structure – The three-dimensional folding pattern, which is super-imposed on the secondary structure, is called the tertiary structure.

Quaternary structure – The quaternary structure is the manner in which two or more independently folded subunits fit together.

F) Nucleic acids – Nucleic acids are long polymers involved in heredity and in the manufacture of different kinds of proteins. The two most important nucleic acids are deoxyribonucleic acid (DNA) and ribonucleic acid (RNA).

Nucleotides – These are the building blocks of nucleic acids. Nucleotides are complex molecules composed of a nitrogenous base, a 5-carbon sugar and a phosphate group.

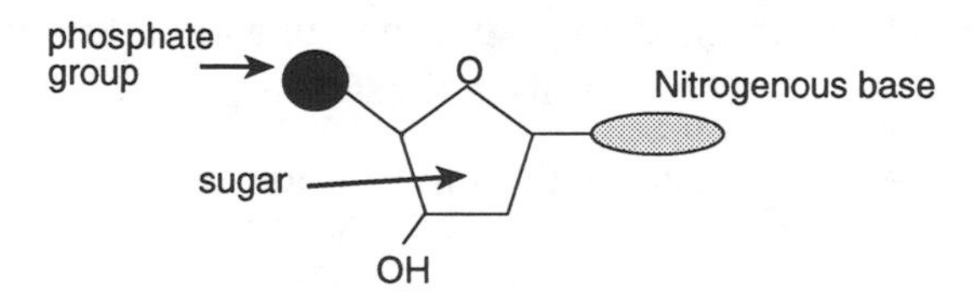

Figure 1.10 Structure of a nucleotide

Deoxyribonucleic Acid (DNA) – Chromosomes and genes are composed mainly of DNA. It is composed of deoxyribose (which is the ribose sugar missing the oxygen on carbon 2), four nitrogenous bases, and phosphate groups.

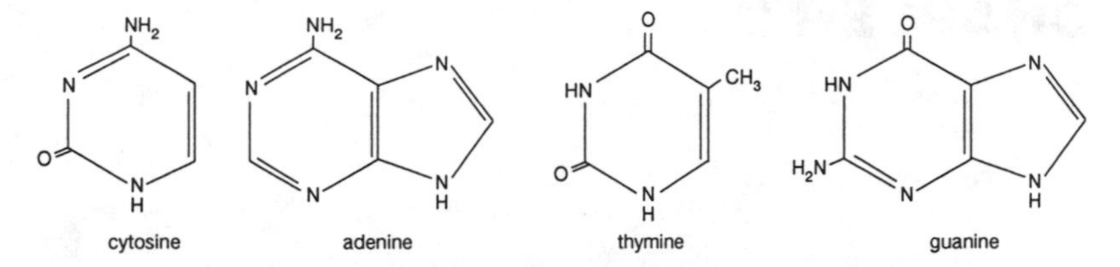

Figure 1.11 The four nitrogenous bases of DNA

Ribonucleic Acid (RNA) – RNA is involved in protein synthesis. It is composed of the sugar ribose and the same nitrogenous base, except uracil is used instead of thymine.

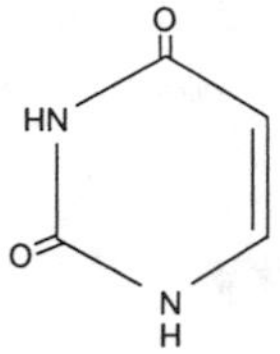

Figure 1.12 Uracil

CHAPTER 2

Cellular Organization

2.1 The Cell Theory

The cell is the unit of structure and the unit of function of most living things. Cells arise from pre-existing cells by reproduction. The biochemical activities that occur within cells depend upon the presence of specific organelles.

Cell Membrane – The cell membrane is a double layer of lipids which surrounds a cell. Proteins are interspersed in this lipid bilayer. The membrane is semi-permeable; it is permeable to water but not to solutes.

The Nucleus – The nucleus is bounded by a pair of membranes. Within the nuclear membrane, there is a semi-fluid medium in which the chromosomes are suspended.

Cytoplasm – The cytoplasm is all the material in a cell located between the nucleus and the plasma membrane. Imbedded in the cytoplasm are the organelles.

2.2 Cellular Organelles

A) Mitochondria – Mitochondria have double-membranes and are the site of chemical reactions that extract energy from foodstuffs and make it available to the cell for all of its energy demanding activities in the form of ATP (adenosine tri-phosphate).

B) Chloroplasts – These are found only in the cells of plants and certain algae. Photosynthesis occurs in the chloroplasts.

C) Plastids – These structures are present only in the cytoplasm of plant cells. The most important plastid, chloroplast, contains chlorophyll, a green pigment.

D) Lysosomes – Lysosomes are membrane-enclosed bodies that function as storage vesicles for many digestive enzymes.

E) Endoplasmic Reticulum (ER) – The endoplasmic reticulum transports substances within the cell. There are two types: smooth and rough.

F) Ribosomes – These organelles are small particles composed chiefly of ribosomal-RNA and are the sites of protein synthesis. Rough ER has ribosomes attached.

G) Golgi Apparatus – The functions of the Golgi apparatus include storage, modification, and packaging of secretory products.

H) Peroxisomes – Peroxisomes are membrane-bound organelles which contain powerful oxidative enzymes.

I) Vacuoles – Vacuoles are membrane-enclosed, fluid-filled spaces. They have their greatest development in plant cells where they store materials such as soluble organic nitrogen compounds, sugars, various organic acids, some proteins, and several pigments.

J) Cell Wall – This is only present in plant cells and is used for protection and support.

K) Centriole and Centrosome – These function in cell division. They are present only in animal cells.

L) Cilia and Flagella – These are hairlike extensions from the cytoplasm of a cell. They both show coordinated beating movements which are the major means of locomotion and ingestion in unicellular organisms.

M) Nucleolus – The nucleolus is a generally oval body composed of protein and RNA. Nucleoli are produced by chromosomes and participate in the process of protein synthesis.

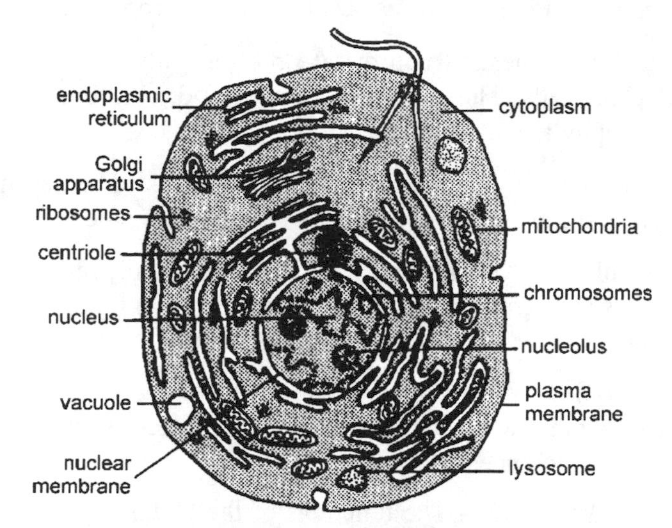

Figure 2.1 Typical Animal Cell

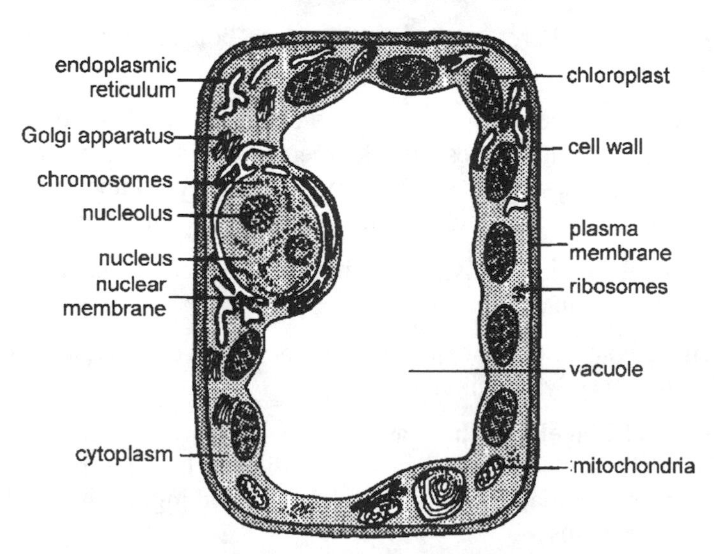

Figure 2.2 Typical Plant Cell

2.3 Prokaryotes vs. Eukaryotes

Prokaryote – Prokaryote refers to bacteria and blue-green algae. Prokaryotic cells have no nuclear membrane, and they lack membrane-bounded subcellular organelles such as mitochondria and chloroplasts. However, the membrane that bounds the cell is folded inward at various points, and carries out many of the enzymatic functions of many internal membranes of eukaryotes.

Eukaryote – Eukaryote refers to all the protists, plants and animals. These are characterized by true nuclei; bounded by a nuclear membrane, and membrane-bound subcellular organelles.

Table 2.1 A comparison of eukaryotic and prokaryotic cells

Characteristic	Eukaryotic cells	Prokaryotic Cells
Chromosomes	multiple, composed of nucleic acids and protein	single, composed only of nucleic acid
Nuclear Membrane	present	absent
Mitochondria	present	absent
Golgi apparatus, endoplasmic reticulum, lysosomes, peroxisomes	present	absent
Photosynthetic apparatus	chlorophyll, when present is contained in chloroplasts	may contain chlorophyll
Microtubules	present	rarely present
Ribosomes	large	small
Flagella	have 9-2 tubular structure	lack 9-2 tubular structure
Cell wall	when present, does not contain muramic acid	contains muramic acid

2.4 Tissues

A tissue is a group of similarly specialized cells which together perform certain special functions.

2.4.1 Plant Tissues

A mature vascular plant possesses several distinct cell types which group together in tissues. The major plant tissues include epidermal, parenchyma, sclerenchyma, chlorenchyma, vascular, and meristematic.

Table 2.2 Summary of Plant Tissues

Tissue	Location	Functions
1. Epidermal	Root	Protection; increases absorption area
	Stem	Protection; reduces H_2O loss
	Leaf	Protection; reduces H_2O loss Regulates gas exchange
2. Parenchyma	Root, stem, leaf	Storage of food and H_2O
3. Sclerenchyma	Stem and leaf	Support
4. Chlorenchyma	Leaf and young stems	Photosynthesis
5. Vascular		
a. Xylem	Root, stem, leaf	Upward transport of fluid
b. Phloem	Leaf, root, stem	Downward transport of fluid
6. Meristematic	Root and stem	Growth; formation of xylem, phloem, and other tissues

2.4.2 Animal Tissues

The cells that make up multicellular organisms become differentiated in many ways. One or more types of differentiated cells are organized into tissues. The basic tissues of a complex animal are the epithelial, connective, nerve, muscle, and blood tissues.

Table 2.3 Summary of Animal Tissues

Tissue	Location	Functions
Epithelial	Covering of body Lining internal organs	Protection Secretion
Muscle		
1. Skeletal	Attached to skeleton bones	Voluntary movement
2. Smooth	Walls of internal organs	Involuntary movement
3. Cardiac	Walls of heart	Pumping blood
Connective		
1. Binding	Covering organs, in tendons and ligaments	Holding tissues and organs together
2. Bone	Skeleton	Support, protection, movement
3. Adipose	Beneath skin and around internal organs	Fat storage, insulation, cushion
4. Cartilage	Ends of bone, part of nose and ears	Reduction of friction, support
Nerve	Brain	Interpretation of impulses, mental activity
	Spinal cord, nerves, ganglions	Carrying impulses to and from all organs
Blood	Blood vessels, heart	Carrying materials to and from cells, carrying oxygen, fighting germs, clotting

2.5 Exchange of Materials Between Cell and Environment

Diffusion – The migration of molecules or ions, as a result of their own random movements, from a region of higher concentration to a region of lower concentration is known as diffusion.

Osmosis – Osmosis is the movement of water through a semipermeable membrane. At constant temperature and pressure, the net movement of water is from the solution with lower concentration to the solution with higher concentration of osmotically active particles.

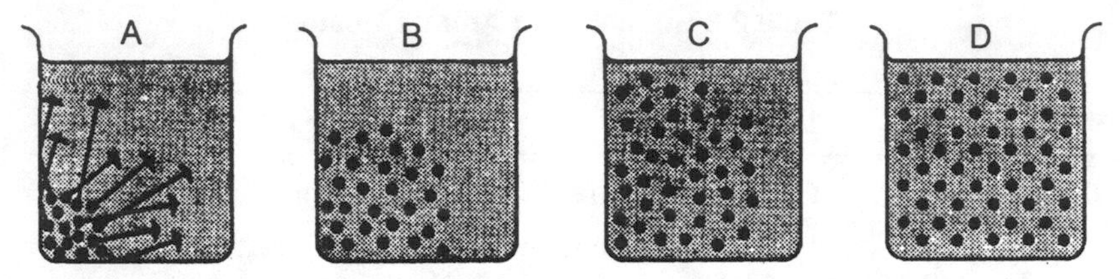

Figure 2.3 The sugar molecules, over a period of time, will be distributed evenly in the water because of diffusion.

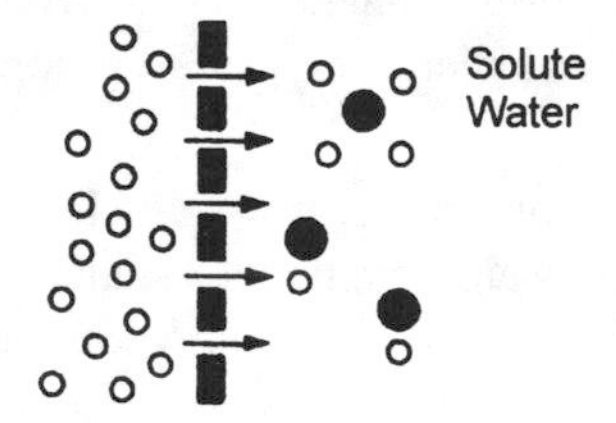

Figure 2.4 The process of osmosis

Active Transport – The movement of ions and molecules against a concentration gradient is referred to as active transport. The cell must expend energy to accomplish the transport. In passive transport, no energy is expended.

Endocytosis – Endocytosis is an active process in which the cell encloses a particle in a membrane-bound vesicle, pinched off from the cell membrane. Endocytosis of solid particles is called phagocytosis.

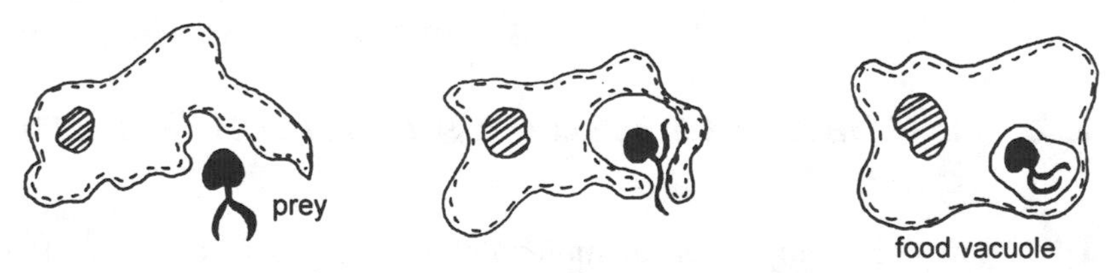

Figure 2.5 Endocytosis in the amoeba

Exocytosis – Exocytosis is the reverse of endocytosis; there is a discharge of vacuole-enclosed materials from a cell by the fusion of the cell membrane with the vacuole membrane.

Isotonic Medium – An isotonic medium is one in which the cell is in osmotic balance because it contains the same concentration of osmotically active particles.

Hypertonic Medium – A hypertonic medium is one in which a cell loses water because the medium contains a higher concentration of osmotically active particles.

Hypotonic Medium – A hypotonic medium is one in which a cell gains water because the medium contains a lower concentration of osmotically active particles.

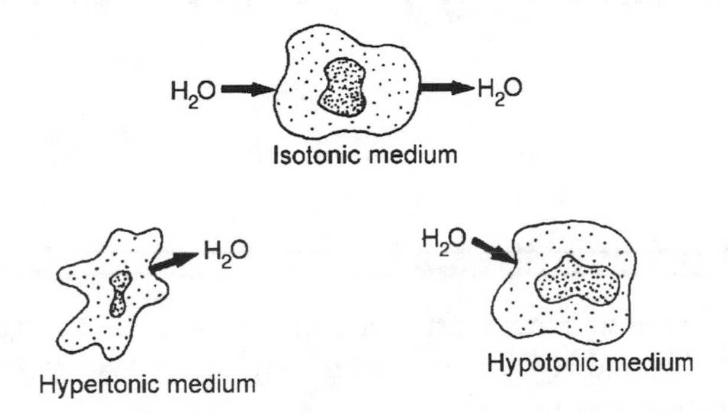

Figure 2.6 Osmotic effects of the fluids bathing cells

CHAPTER 3

Cellular Metabolism and Energy Pathways

3.1 Photosynthesis

Photosynthesis is the basic food-making process through which inorganic CO_2 and H_2O are transformed to organic compounds, specifically carbohydrates.

Chloroplasts absorb light energy and use CO_2 and H_2O to synthesize carbohydrates. Oxygen, which is formed as a byproduct, is either eliminated into the air through the stomates, stored temporarily in the air spaces, or used in cellular respiration.

An overall chemical description of photosynthesis is the equation

$$6\,CO_2 + 6\,H_2O \xrightarrow[\text{chlorophyll}]{\text{light}} C_6H_{12}O_6 + 6\,O_2$$

3.1.1 Light Reaction (Photolysis)

A first step in photosynthesis is the decomposition of water molecules to separate hydrogen and oxygen components. This decomposition is associated with processes involving chlorophyll and light and is thus known as the light reaction.

3.1.2 Dark Reaction (CO_2 Fixation)

In this second phase, the hydrogen that results from photolysis reacts with CO_2 and carbohydrate forms. CO_2 fixation does not require light.

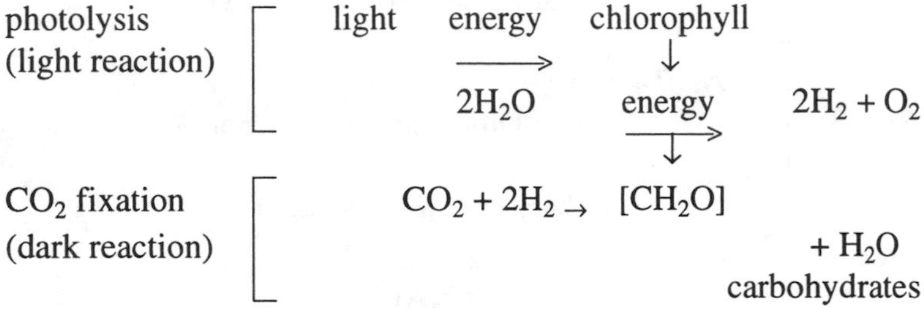

photolysis (light reaction)

$$2H_2O \xrightarrow{\text{light energy}} 2H_2 + O_2 \quad (\text{chlorophyll} \downarrow \text{energy})$$

CO_2 fixation (dark reaction)

$$CO_2 + 2H_2 \rightarrow [CH_2O] + H_2O$$

carbohydrates

Figure 3.1 Photolysis and CO_2 fixation

3.2 Cellular Respiration

During cellular respiration, glucose must first be activated before it can break down and release energy. After activation, glucose enters into numerous reactions occurring in two stages. One stage is the anaerobic phase of cellular respiration and the other is the aerobic phase.

A) Glycolysis – Glycolysis refers to the breakdown of glucose which marks the start of the anaerobic reactions of cellular respiration. ATP is the energy source which activates glucose and initiates the process of glycolysis.

The steps in Figure 3.2 are summarized as follows:

Step 1 – Activation of glucose

Step 2 – Formation of sugar diphosphate

Step 3 – Formation and oxidation of PGAL, phosphoglyceraldehyde

Step 4 – Formation of pyruvic acid ($C_3H_4O_3$)
Net gain of two ATP molecules

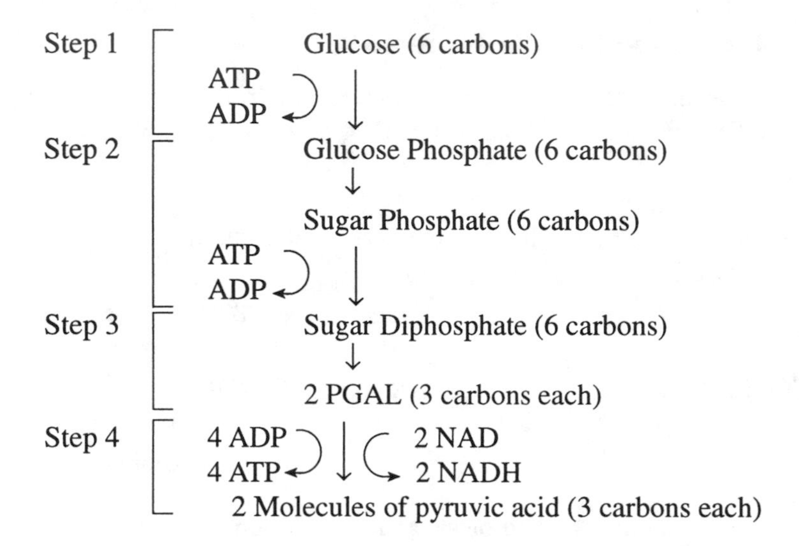

Figure 3.2 The Major Steps in Glycolysis

B) Krebs Cycle (Citric Acid Cycle) – The Krebs cycle is the final common pathway by which the carbon chains of amino acids, fatty acids, and carbohydrates are metabolized to yield CO_2. Pyruvic acid is converted to acetyl coenzyme A and, through a series of reactions, citric acid is formed. (See figure on following page.)

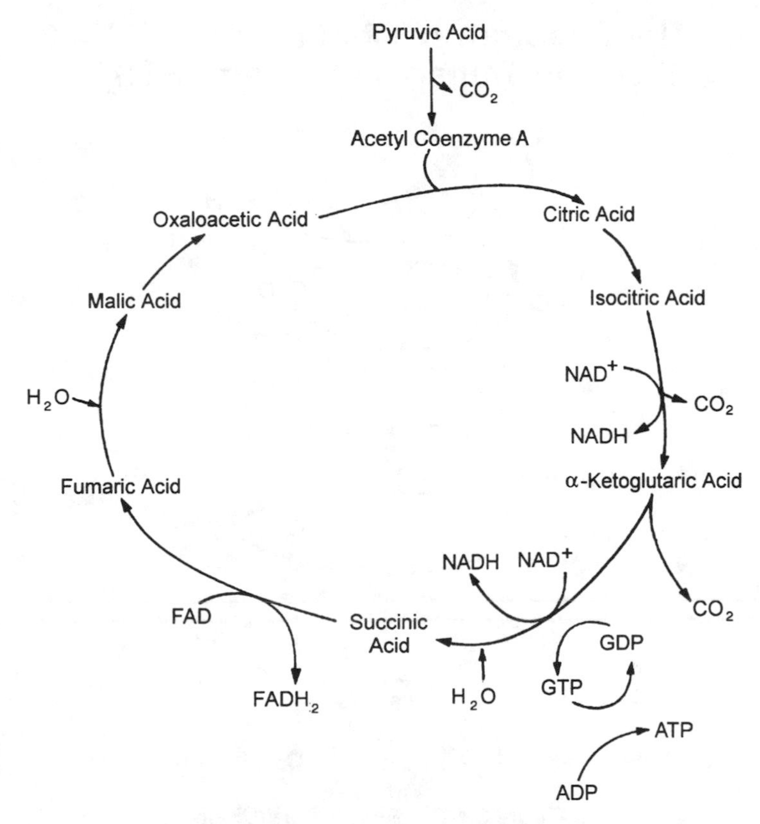

Figure 3.3 Summary of the Krebs Cycle

3.3 ATP and NAD

ATP – ATP stands for adenosine triphosphate and is a coenzyme essential for the breakdown of glucose. When the bonds in ATP are hydrolyzed, a large amount of energy is released. It takes 7 Kcal to phosphorylate (add a phosphate) to ADP to make ATP.

NAD – NAD stands for nicotinamide adenine dinucleotide. Like ATP, it is also a coenzyme. NAD participates in a large number of oxidation-reduction reactions in cells, including those in cellular respiration. It is an electron acceptor and donator. Oxidation of NADH releases 53 Kcal.

3.4 The Respiratory Chain (Electron Transport System [ETS])

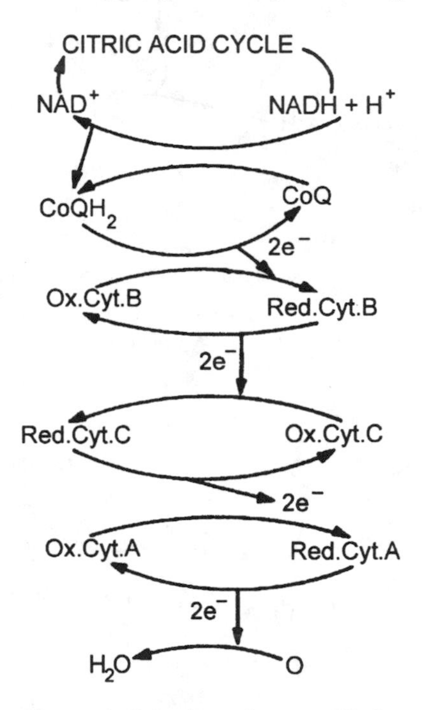

Figure 3.4 The Respiratory Chain

This system consists of a series of enzymes and coenzymes which pick up, hold, and then transfer hydrogen atoms among themselves until the hydrogen reaches its final acceptor, which is oxygen. Cytochromes are the enzymes and coenzymes involved in transferring hydrogen.

The cytochromes, together with other enzymes, split hydrogen atoms attached to compounds such as $NADH_2$, into hydrogen ions and electrons. Each cytochrome then passes the hydrogen ions and electrons to another cytochrome in the series.

3.5 Enzymes

Enzymes are protein catalysts that lower the amount of activation energy needed for a reaction, allowing it to occur more rapidly. The enzyme binds with the substrate but resumes its original conformation after completion of the reaction.

Substrate – A substrate is the molecule upon which an enzyme acts. The atoms of the substrate are arranged so as to fit into the active site of the enzyme that acts upon it. An enzyme usually affects only one substrate.

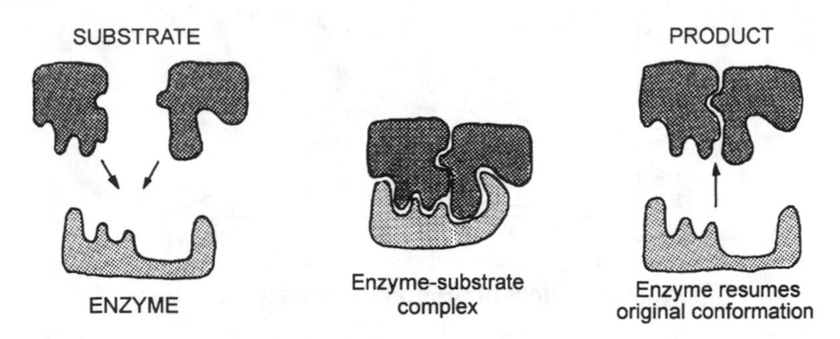

Figure 3.5 The action of an enzyme

Allosteric Enzyme – An allosteric enzyme is one that can exist in two distinct conformations. Usually, the enzyme is active in one conformation and inactive in the other.

Allosteric Inhibition – During the process of allosteric inhibition, an inhibitory molecule called a negative modulator binds to the enzyme and stabilizes it in its inactive conformation, preventing the reaction from proceeding.

Coenzymes – These are metal ions or non-proteinaceous organic molecules that bind briefly and loosely to some enzymes. The coenzyme is necessary for the catalytic reaction of such enzymes.

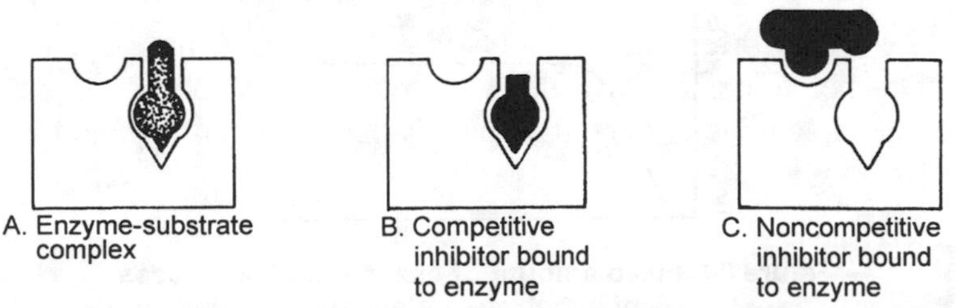

Figure 3.6 A. Enzyme-substrate complex; B. Competitive inhibition; C. Non-competitive inhibition

Factors influencing the rate of enzyme action:

1. pH
2. temperature
3. concentration of enzyme and substrate

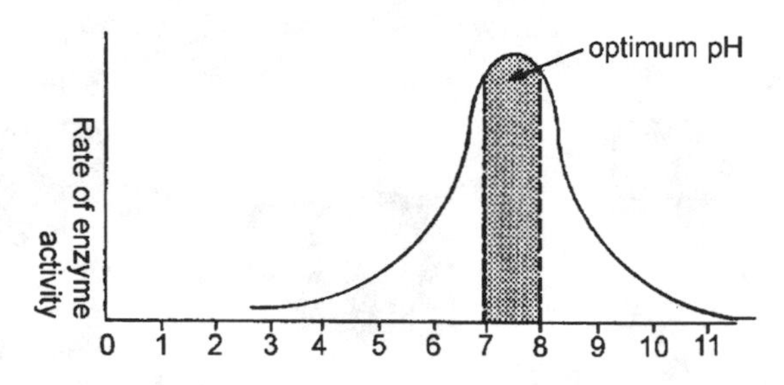

Figure 3.7 Effect of pH on rate of enzyme action

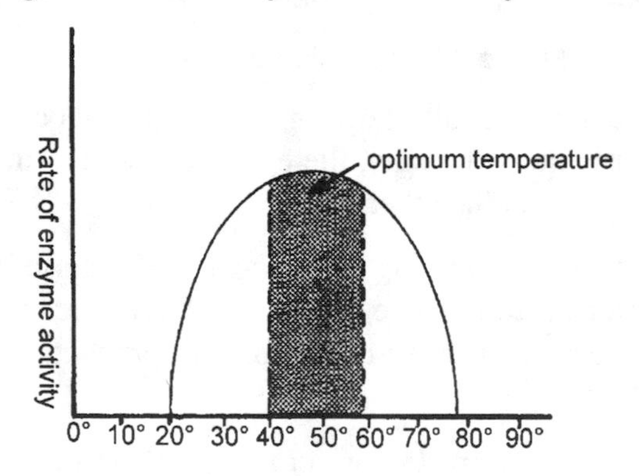

Figure 3.8 Effect of temperature upon rate of enzyme activity

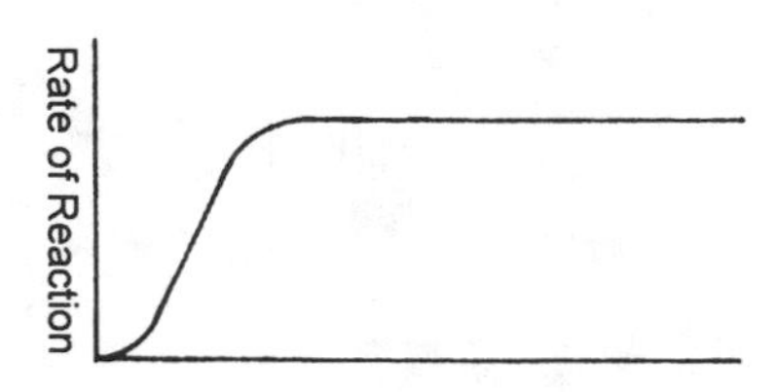

Figure 3.9 Fixed amount of enzyme and an excess of substrate molecules

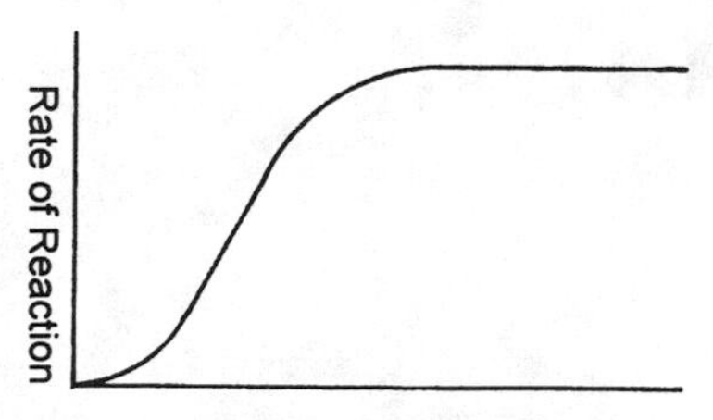

Figure 3.10 Fixed number of substrate molecules and an excess of enzyme molecules

CHAPTER 4

Nutrition in Plants and Animals

4.1 Importance of Digestion

After food has been ingested, it must be digested or broken down into smaller molecules so that these molecules can pass through plasma membranes and reach the cytoplasm inside the cells.

Carbohydrate molecules must be broken down to small molecules of simple sugars; protein molecules into small molecules of amino acids; and lipid molecules must be broken down into molecules of fatty acids and glycerol.

Intracellular Digestion – In some heterotrophic organisms, digestion occurs after the solid material has actually been engulfed by a cell. This is common in the protozoans.

Extracellular Digestion – Organisms carry out digestion outside their cells, usually within the cavity of a digestive system; this is referred to as extracellular digestion. The simplest approach to this is employed by saprophytes who secure their food from nonliving but organic matter. This is the only type of digestion in many multicellular organisms, including humans.

In higher organisms, extracellular digestion takes place in two phases, mechanical and chemical.

Mechanical Phase of Digestion

Chewing and grinding break down food into smaller particles so that the total surface area of the food exposed to chemical action is increased.

Chemical Phase of Digestion

Chemical digestion which is carried on by certain enzymes reduces the size of the particles. By means of hydrolysis reactions, enzymes break the chemical bonds that hold the large molecules together.

4.2 The Human Digestive System

The digestive system of humans consists of the alimentary canal and several glands. This alimentary canal consists of the oral cavity (mouth), pharynx, esophagus, stomach, small intestine, large intestine, and the rectum.

A) Oral cavity (mouth) – The mouth cavity is supported by jaws and is bound on the sides by the teeth, gums, and cheeks. The tongue binds the bottom and the palate binds the top. Food is pushed between the teeth by the action of the tongue so it can be chewed and swallowed. Saliva is the digestive juice secreted that begins the chemical phase of digestion.

B) Pharynx – Food passes from the mouth cavity into the pharynx where the digestive and respiratory passages cross. Once food passes the upper part of the pharynx, swallowing becomes involuntary.

C) Esophagus – Whenever food reaches the lower part of the pharynx, it enters the esophagus and peristalsis pushes the food further down the esophagus into the stomach.

D) Stomach – The stomach has two muscular valves at both ends: the cardiac sphincter which controls the passage of food from the esophagus into the stomach and the pyloric sphincter which is re-

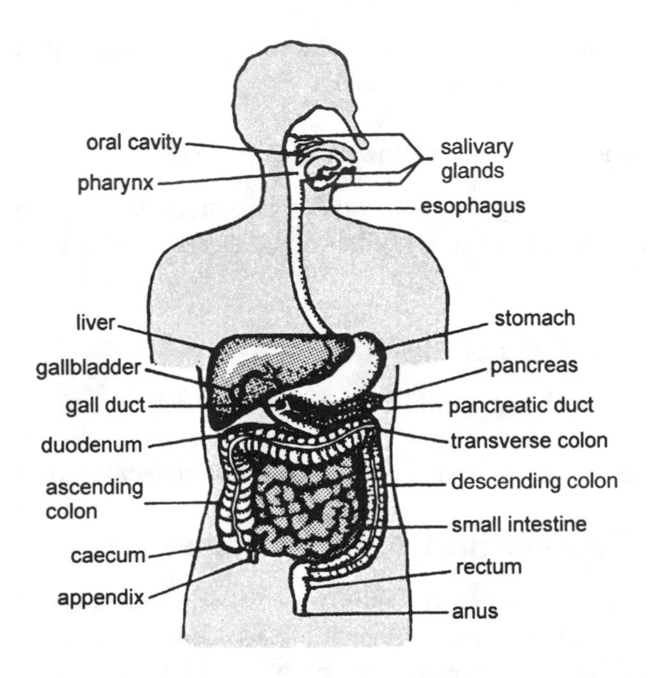

Figure 4.1 Human digestive system (The organs are slightly displaced, and the small intestine is greatly shortened.)

sponsible for the control of the passage of partially digested food from the stomach to the small intestine. Gastric juice is also secreted by the gastric glands lining the stomach walls. Gastric juice begins the digestion of proteins.

E) Pancreas – The pancreas is the gland formed by the duodenum and the under surface of the stomach. It is responsible for producing pancreatic fluid which aids in digestion. Sodium bicarbonate, amylase, lipase, trypsin, chymotrypsin, carboxypeptidase, and nucleases are all found in the pancreatic fluid.

F) Small Intestine – The small intestine is a narrow tube between 20 and 25 feet long divided into three sections: the duodenum, jejunum, and ileum. The final digestion and absorption of disaccharides, peptides, fatty acids, and monoglycerides is the work of villi which line the small intestine.

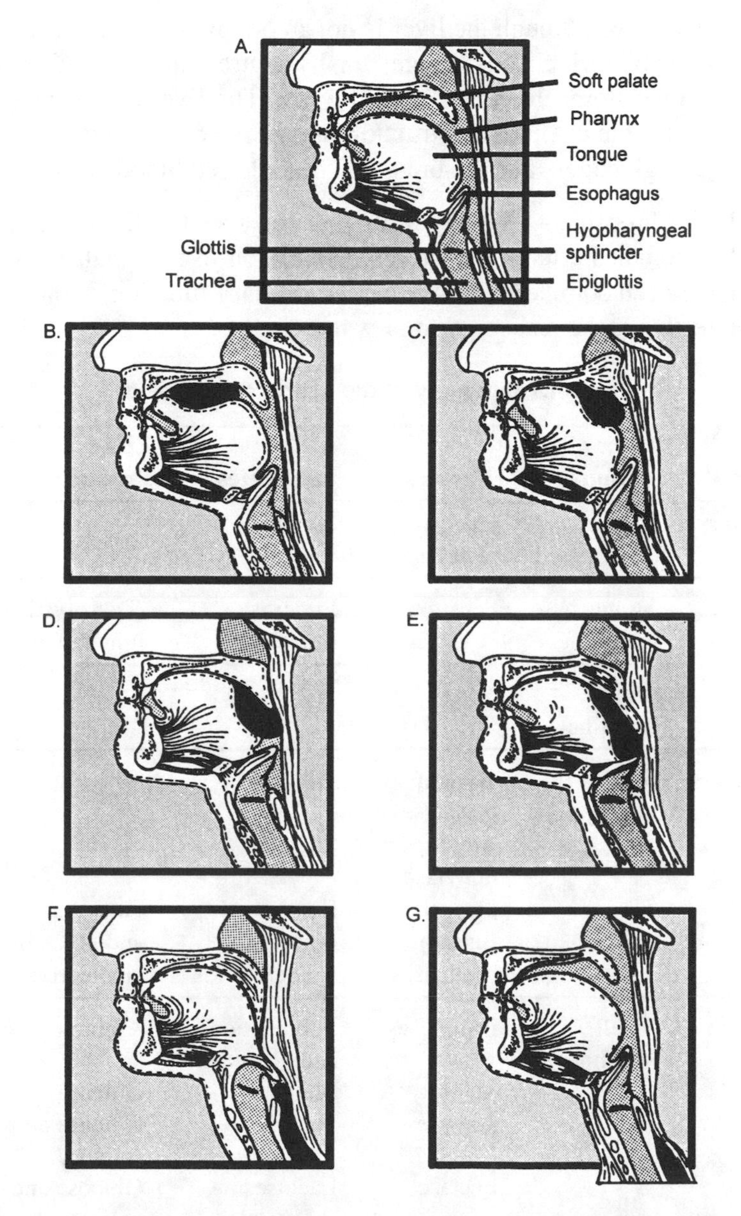

Figure 4.2 Movement of a bolus of food through the pharynx and upper esophagus during swallowing A) Mouth and pharynx at rest B) Early oral phase C) Late oral phase D) Early pharyngeal phase E) Middle pharyngeal phase F) Late pharyngeal phase G) Esophageal phase

G) Liver – Even though the liver is not an organ of digestion, it does secrete bile which aids in digestion by neutralizing the acid chyme from the stomach and emulsifying fats. The liver is also responsible for the chemical destruction of excess amino acids, the storage of glycogen, and the breakdown of old red blood cells.

H) Large Intestine – The large intestine receives the liquid material that remains after digestion and absorption in the small intestine have been completed. However, the primary function of the large intestine is the reabsorption of water.

Table 4.1 Summary of the action of enzymes

Gland	Place of Action	Enzymes	Substrates	End Products
Salivary	Mouth	Ptyalin (amylase)	Starch	
Gastric	Stomach	Pepsin	Proteins (minerals)	(Dissolved minerals)
Liver	Small intestine	None		
Pancreas	Small intestine	Trypsin (protease)	Proteins	
		Amylopsin (amylase)	Starch	
		Steapsin (lipase)	Emulsified lipids	Fatty acids and glycerol
		Nucleases	Nucleic acids	Nucleotides
Intestinal	Small intestine	Peptidases	Polypeptides and dipeptides	Amino acids
		Maltase	Maltose	Glucose
		Sucrase	Sucrose	Glucose and fructose
		Lactase	Lactose	Glucose and galactose

4.3 Ingestion and Digestion in Other Organisms

A) Hydra – The hydra possesses tentacles which have stinging cells (nematocysts) which shoot out a poison to paralyze the prey. If successful in capturing an animal, the tentacles push it into the hydra's mouth. From there, the food enters the gastric cavity. The hydra uses both intracellular and extracellular digestion.

B) Earthworm – As the earthworm moves through soil, the suction action of the pharynx draws material into the mouth cavity. Then from the mouth, food goes into the pharynx, the esophagus, and then the crop which is a temporary storage area. This food then passes into a muscular gizzard where it is ground and churned. The food mass finally passes into the intestine; any undigested material is eliminated through the anus.

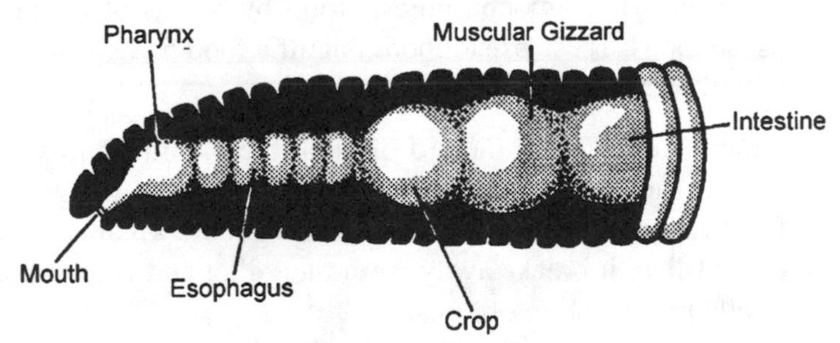

Figure 4.3 The digestive system of the earthworm

C) Grasshopper – The grasshopper is capable of consuming large amounts of plant leaves. This plant material must first pass through the esophagus into the crop, a temporary storage organ. It then travels to the muscular gizzard where food is ground. Digestion takes place in the stomach. Enzymes secreted by six gastric glands are responsible for digestion. Absorption takes place mainly in the stomach. Undigested material passes into the intestine, collects in the rectum, and is eliminated through the anus.

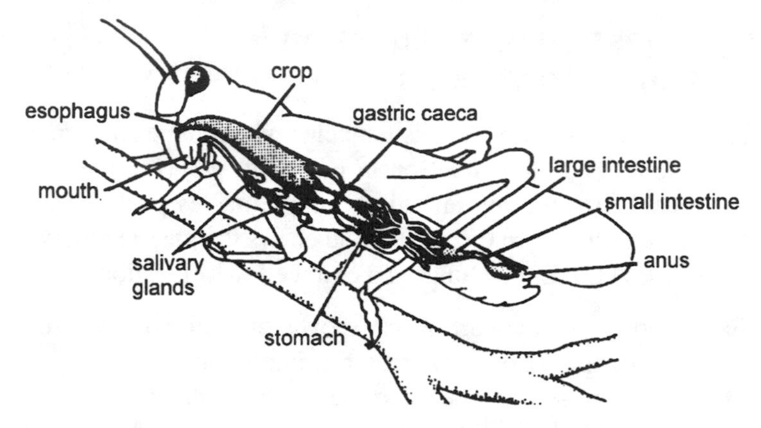

Figure 4.4 The digestive system of the grasshopper

D) Protozoa

1) Amoeba – The amoeba ingests food by means of temporary pseudopods. These pseudopods engulf a food particle and form a food vacuole within the cytoplasm.

2) Paramecium – With the aid of cilia, the paramecium sweeps food particles into the oral groove and then into the gullet. A food vacuole forms at the end of the gullet; when this vacuole is filled, it breaks away from the gullet and is transported to other parts of the organism.

4.4 Transport of Food in Vascular Plants

A) Leaf – The leaf consists of an upper epidermis, mesophyll layer, and lower epidermis. Its primary function is to change inorganic substances to organic substances by the process of photosynthesis. A leaf also functions in the exchange of gases between the plant and the atmosphere.

The veins in the leaf, known as xylem and phloem tubes, transport fluid materials in the leaf.

1) Xylem – carries water and dissolved minerals from the stem and roots to the leaf cells.

2) Phloem – transports materials from the leaf to the stem and roots.

B) Stem – The stem consists of an epidermis, sclerenchyma, parenchyma, and conducting tissue which are the xylem and phloem. One of the primary functions of the stem is to transport raw materials from the roots to the leaves and manufactured products to the roots and other plant organs.

C) Root – The water that is needed by the plants enters by way of the roots. Water and dissolved minerals diffuse into the root hairs and pass through the cortex cells to the cells of the xylem.

CHAPTER 5

Gas Exchange in Plants and Animals

5.1 Respiration in Humans

The respiratory system in humans begins as a passageway in the nose. Inhaled air then passes through the pharynx, trachea, bronchi, and lungs.

A) Nose – The nose is better adapted to inhale air than the mouth. The nostrils, the two openings in the nose, lead into the nasal passages which are lined by the mucous membrane. Just beneath the mucous membrane are capillaries which warm the air before it reaches the lungs.

B) Pharynx – Air passes via the nasal cavities to the pharynx where the paths of the digestive and respiratory systems cross.

C) Trachea – The upper part of the trachea, or windpipe, is known as the larynx. The glottis is the opening in the larynx; the epiglottis, which is located above the glottis, prevents food from entering the glottis and obstructing the passage of air.

D) Bronchi – The trachea divides into two branches called the bronchi. Each bronchus leads into a lung.

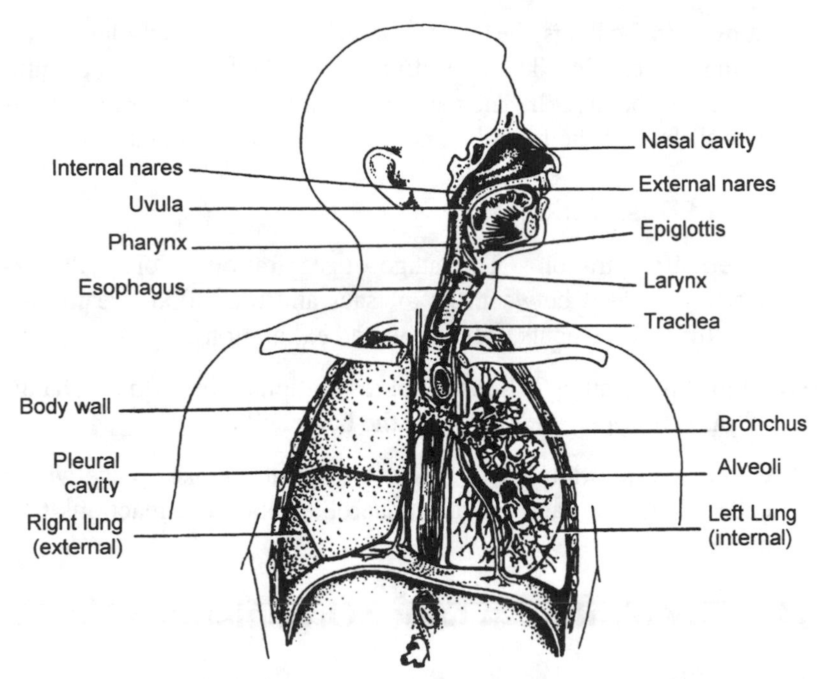

Figure 5.1 Diagram of the human respiratory system

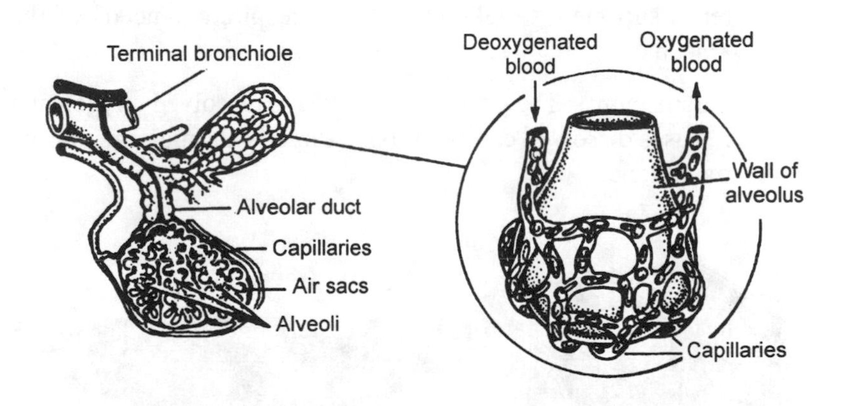

Figure 5.2 Diagram of a small portion of the lung, highly magnified, showing the air sacs at the end of the alveolar ducts, the alveoli in the walls of the air sacs, and the proximity of the alveoli and the pulmonary capillaries containing red blood cells

E) Lungs – In the lungs, the bronchi branch into smaller tubules known as the bronchioles. The finer divisions of the bronchioles eventually enter the alveoli. The cells of the alveoli are the true respiratory surface of the lung. It is here that gas exchange takes place.

Stages of Respiration

A) External respiration – This stage of respiration involves the exchange of gases between the air sacs and the blood stream, and breathing movements (inhalation and exhalation).

B) Internal respiration – This stage of respiration involves the exchange of gases between the blood and the body cells.

C) Cellular respiration – This stage of respiration takes place within cells where both aerobic and anaerobic oxidation reactions take place.

5.2 Respiration in Other Organisms

A) Protozoa

1) Amoeba – Simple diffusion of gases between the cell and water is sufficient to take care of the respiratory needs of the amoeba.

2) Paramecium – The paramecium takes in dissolved oxygen and releases dissolved carbon dioxide directly through the plasma membrane.

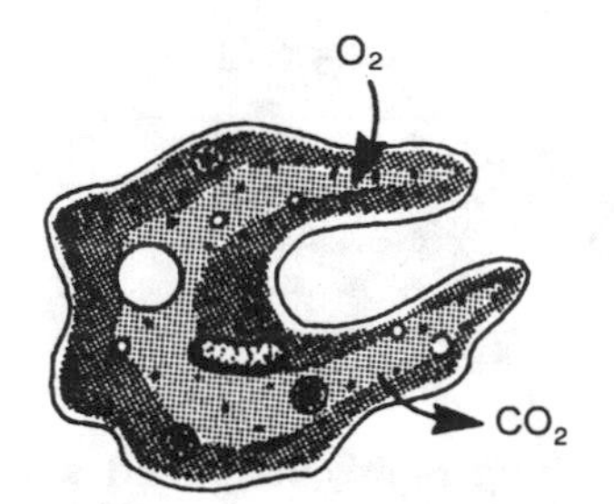

Figure 5.3 Respiration in the amoeba

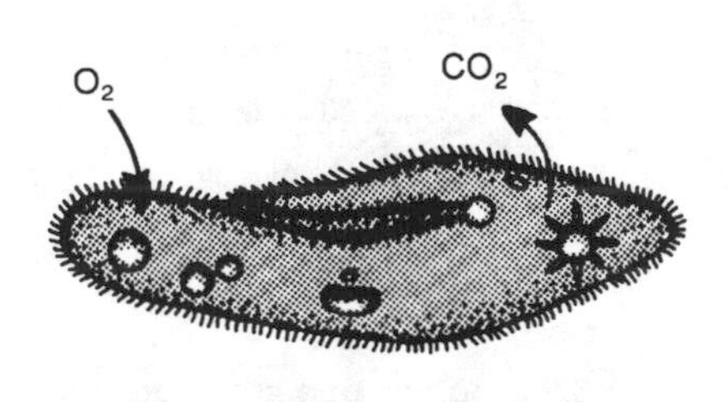

Figure 5.4 Respiration in the paramecium

B) Hydra – Dissolved oxygen and carbon dioxide diffuse in and out of two cell layers through the plasma membrane.

C) Grasshopper – The grasshopper carries on respiration by means of spiracles and tracheae. Blood plays no role in transporting oxygen and carbon dioxide. Muscles of the abdomen pump air into and out of the spiracles and the tracheae.

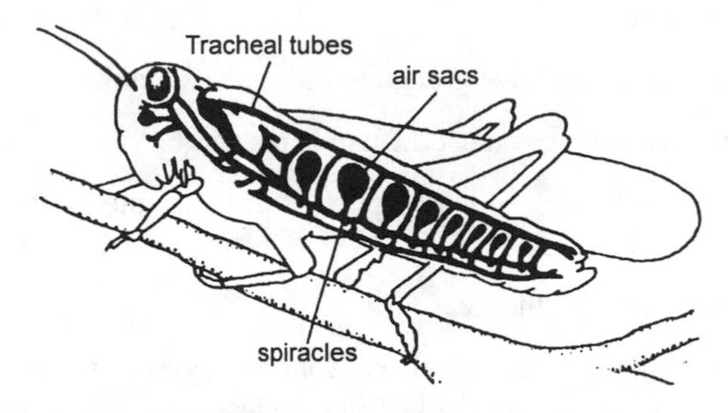

Figure 5.5 Respiration in the grasshopper

D) Earthworm – The skin of the earthworm is its respiratory surface. Oxygen from the air diffuses into the capillaries of the skin and joins with hemoglobin dissolved in the blood plasma. This oxyhemoglobin is released to the tissue cells. Carbon dioxide from the tissue cells diffuses into the blood. When the blood reaches the capillaries in the skin again, the carbon dioxide diffuses through the skin into the air.

Table 5.1
Comparison of various respiratory surfaces among organisms

Organism	Respiratory Surface
Protozoan	Plasma membrane
Hydra	Plasma membrane of each cell
Grasshopper	Tracheae network
Earthworm	Moist skin
Human	Air sacs in lungs

5.3 Gas Exchange in Plants

A) Gas Exchange in Roots and Stems

Plants are able to exist without specialized organs for gas exchange because:

1) there is little transport of gases from one part of the plant to another.

2) plants do not have great need for gas exchange.

3) the distance gas must diffuse is not large.

4) most cells have at least a portion of their outer surface in contact with air.

B) Gas Exchange in the Leaf

The exchange of gases in the leaf for photosynthesis occurs through pores in the surface of the leaf known as stomata. Usually, the stomata open during the presence of light and close in its absence. The most direct cause of this is the change of turgor in the guard cells.

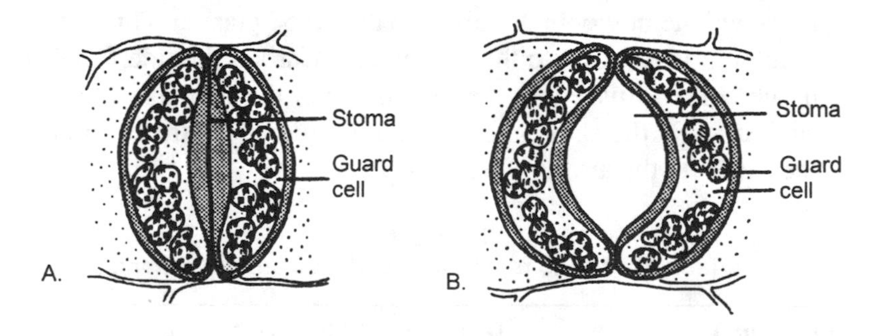

Figure 5.6 A. Stomata closed
B. Stomata open when turgor builds in guard cells

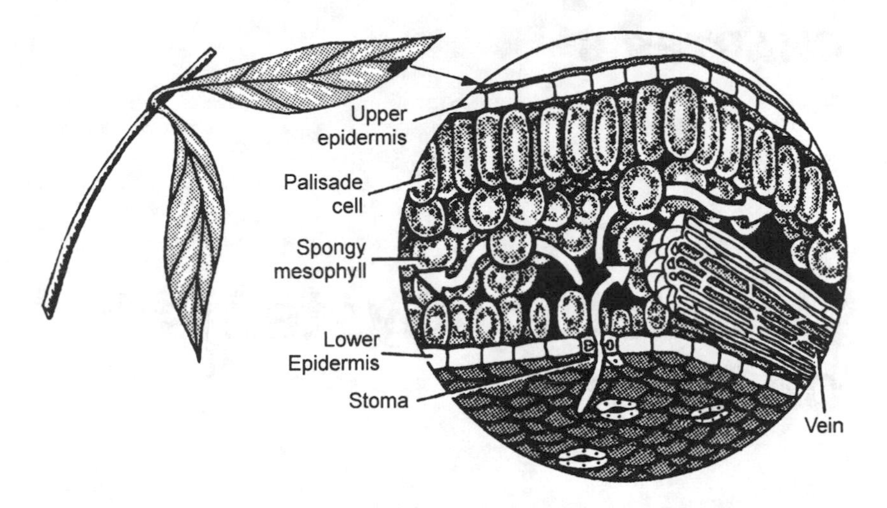

Figure 5.7 A diagram representing loosely arranged cells in a leaf which allow for rapid diffusion of gas

CHAPTER 6

Circulatory Systems of Animals

6.1 The Human Circulatory System

Humans have a closed circulatory system in which the blood moves entirely within the blood vessels. This circulatory system consists of the heart, blood, veins, arteries, capillaries, lymph, and lymph vessels.

A) Heart

The heart is a pump covered by a protective membrane known as the pericardium and divided into four chambers. These chambers are the left and right atria, and the left and right ventricles.

1) Atria – The atria are the upper chambers of the heart that receive blood from the superior and inferior vena cava. This blood is then pumped to the lower chambers or ventricles.

2) Ventricles – The ventricles have thick walls as compared to the thin walls of the atria. They must pump blood out of the heart to the lungs and other distant parts of the body.

The heart also contains many important valves. The tricuspid valve is located between the right atrium and the right ventricle. It prevents the backflow of blood into the atrium after the contraction of the right ventricle. The bicuspid valve, or mitral valve, is situated between the left atrium and the left ventricle. It prevents the backflow of blood into the left atrium after the left ventricle contracts.

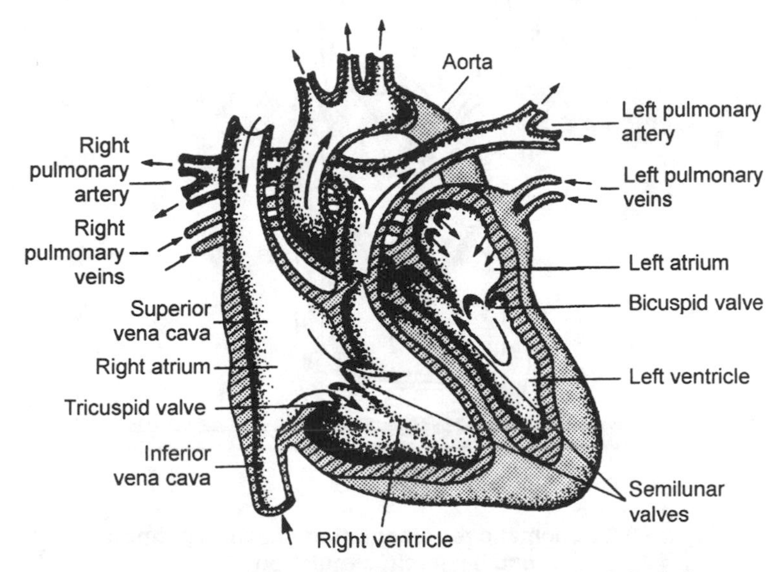

Figure 6.1 The functional parts of the heart and direction of blood flow through its chambers

6.2 Blood Circuits

The pulmonary artery carries blood from the heart into the lungs to oxygenate it. The branches of the aorta carry blood to the other systems of the body. Semilunar valves, at the base of these arteries, prevent the backflow of blood into the ventricles.

A) Pulmonary circulation – The blood circuit from the heart to the lungs and then back to the heart is called pulmonary circulation.

B) Systemic circulation – In this blood circuit, blood circulates from the left ventricle into the aorta and then to all body systems, except the lungs, and back to the heart. There are three branches of systemic circulation. These are the coronary circulation, the hepatic portal system, and renal circulation.

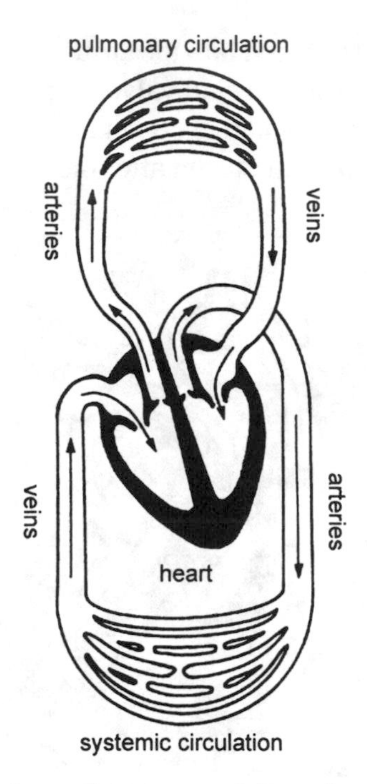

Figure 6.2 Schematic representation showing pulmonary and systemic circulation

C) Veins – All the veins in the body carry blood to the heart. Veins possess small amounts of muscle tissue in their walls and, therefore, the walls are thinner. Valves, which prevent the backflow of blood, are found along the length of the vein. Every vein in the body, except the pulmonary vein, carries deoxygenated blood.

D) Arteries – Every artery in the body, except the pulmonary artery, carries oxygen rich blood. Every artery in the body carries blood away from the heart. Unlike the thin walls of veins, arteries possess relatively thick walls.

E) Capillaries – The capillaries are the most abundant blood vessels in the body. They connect arterioles and venules to complete the circuit of blood. Capillaries function in the exchange of material between the cells and the blood. The walls of the capillary consist of a single layer of flattened epithelial cells.

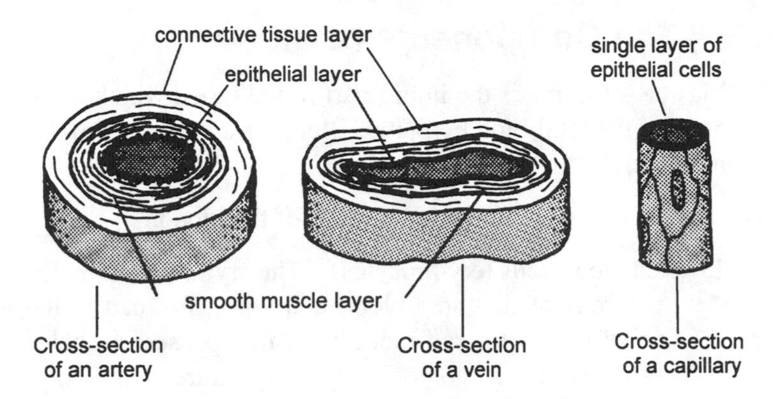

Figure 6.3 Comparison of blood vessels

F) Lymph – Lymph is the intercellular fluid present around every internal body cell. It is composed of water and the dissolved substances that pass out of the capillaries and the tissue cells. Lymph is the medium through which materials are exchanged between the tissue cells and blood vessels.

G) Lymph vessels – The lymphatic vessels function to return fluid from the tissue spaces to the circulation. Lymphatic vessels join larger collecting lymphatics such as the thoracic duct. The thoracic duct carries lymph upward across the chest and empties into a large vein near the base of the neck. Lymph travels in only one direction, from the body organs to the heart.

1) Lymph nodes

Lymph nodes are the thousands of pea-shaped structures lining the lymph vessels. They filter out bacteria from the lymph. The white blood cells produced by the lymph nodes destroy bacteria.

2) Spleen

The spleen is a sac-like mass of lymphatic tissue located near the stomach. The blood that passes into it is filtered. The spleen stores red blood cells so that when bleeding occurs, it contracts and forces the stored red blood cells into circulation.

6.3 The Components of Blood

A) Plasma – Plasma is the liquid part of the blood which constitutes 55% of the total blood volume. Blood plasma is essential for homeostasis.

B) Blood Cells – These constitute 45% of the blood.

1) Red blood cells (erythrocytes) – The erythrocytes are the most numerous of the three blood cell types. Formed in the marrow of bones, the red blood cells first possess a nucleus and not very much hemoglobin. As they mature, hemoglobin constitutes 90% of the dry cell weight. The chief function of erythrocytes is to carry oxygen to all parts of the body and to remove some carbon dioxide.

2) White blood cells (leukocytes) – A white blood cell has a nucleus but lacks hemoglobin. They are larger than red blood cells and generally function to protect the body against disease. There are several types of white blood cells. These are the neutrophils, monocytes, lymphocytes, eosinophils, and basophils. Some are formed in certain bones, others, in the lymph nodes.

3) Platelets – These are cell fragments which are produced by large cells in the bone marrow. Platelets are much smaller than red or white blood cells. They play an important role in blood clotting.

6.4 The Functions of Blood

Blood:

A) Transports materials to and from all the tissues of the body.

B) Defends the body against infectious diseases.

6.5 The Clotting of Blood

The sequence of events in blood clotting is initiated with the breakdown of platelets. With this, thromboplastin is liberated. In the presence of excess thromboplastin and calcium ions, the clotting enzyme prothrombin is changed to thrombin. This thrombin then acts on fibrinogen, a dissolved protein, and converts it into fibrin. At the wound site, the fibrin serves to trap the blood cells. A clot is soon formed which stops loss of blood from the severed vessel.

1) Platelets + Damaged Cells $\xrightarrow[\text{in Blood}]{\text{Clotting Factors}}$ Thromboplastin

2) Prothrombin $\xrightarrow[Ca^{++}]{\text{Thromboplastin}}$ Thrombin

3) Fibrinogen $\xrightarrow{\text{Thrombin}}$ Fibrin

4) Fibrin + Cells + Dried Serum $\longrightarrow$ Blood Clot

Figure 6.4 The sequence in blood clotting

6.6 Transport Mechanisms in Other Organisms

A) Protozoans

Most protozoans are continually bathed by food and oxygen because they live in water or another type of fluid. With the process of cyclosis or diffusion, digested materials and oxygen are distributed within the cell, and water and carbon dioxide are removed. Proteins are transported by the endoplasmic reticulum.

B) Hydra

Like the protozoans, materials in the hydra are distributed to the necessary organelles by diffusion, cyclosis, and by the endoplasmic reticulum.

C) Earthworm

The circulatory system of the earthworm is known as a "closed" system because the blood is confined to the blood vessels at all times. A pump that forces blood to the capillaries consists of five pairs of aortic loops. Contraction of these loops forces blood into the ventral blood vessel. This ventral blood vessel transports blood toward the rear of the worm. The dorsal blood vessel forces blood back to the aortic loops at the anterior end of the worm.

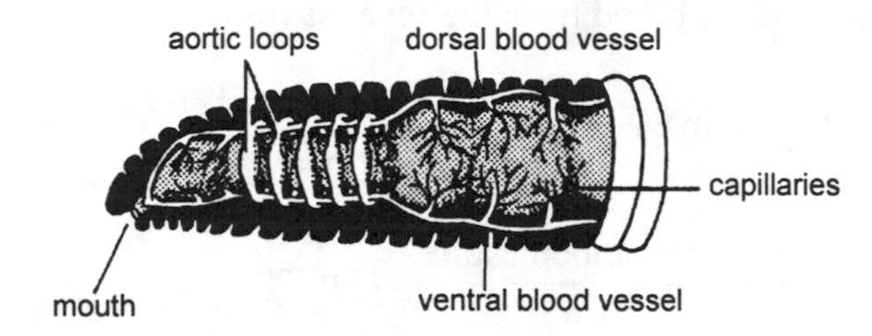

Figure 6.5 "Closed" circulatory system of the earthworm

D) Grasshopper

The grasshopper possesses an "open" circulatory system where the blood is confined to vessels during only a small portion of its circuit through the body. The blood is pumped by the contractions of a tubular heart and a short aorta with an open end. Blood from the heart flows into the aorta and then into sinuses. The blood then returns to the heart.

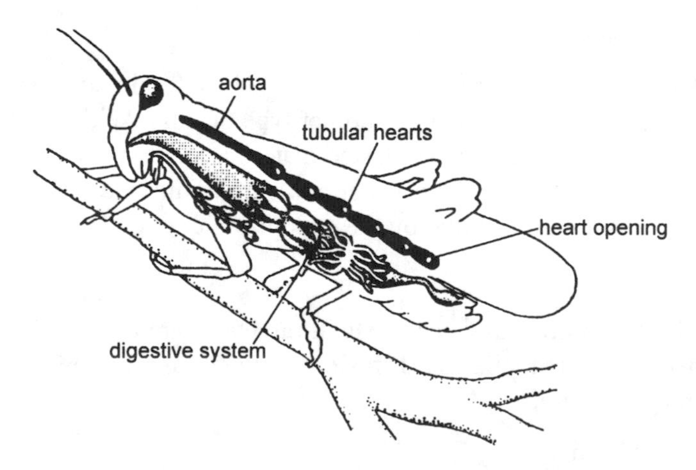

Figure 6.6 "Open" circulatory system of the grasshopper

Table 6.1
Summary of types of circulation present in certain organisms

Organism	System of Circulation
Protozoan	No system. Cyclosis, endoplasmic reticulum, and diffusion distribute materials.
Hydra	No system. Same as protozoan.
Earthworm	Closed circulatory system. Has "hearts," arteries, veins, capillaries.
Grasshopper	Open circulatory system. Has heart, aorta, sinuses.
Human	Closed circulatory system. Has heart, arteries, veins, capillaries, and lymph vessels.

CHAPTER 7

Excretion and Homeostasis

7.1 Excretion in Plants

Plants lack specific organs of excretion and reuse most of the wastes that they produce.

A) Catabolism in plants is usually much lower than in animals; therefore, metabolic wastes accumulate at a slower rate.

B) Green plants use much of the waste products produced by catabolism in anabolic processes.

7.2 Excretion in Humans

The organs of excretion which remove metabolic wastes from a cell or organ in humans include the skin, lungs, liver, the kidneys, and the large intestine.

A) Skin – The skin functions as an organ of excretion as well as protection against injury and regulation of body fluids. The sweat glands of the skin remove water, mineral salts, and urea from the blood.

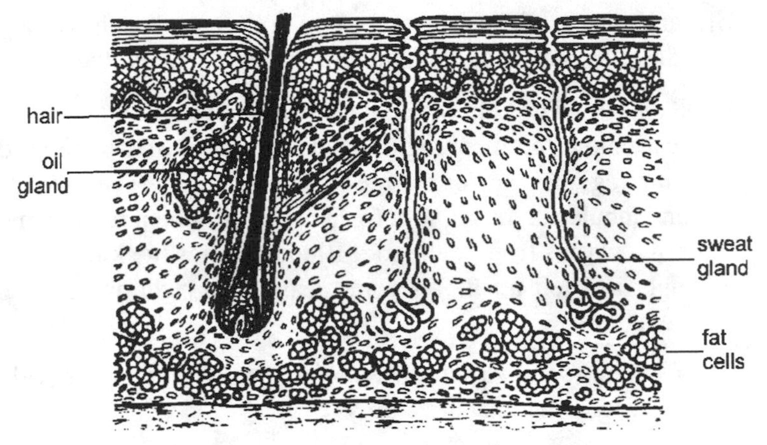

Figure 7.1 Section of the human skin

B) Lungs – The lungs excrete water and carbon dioxide.

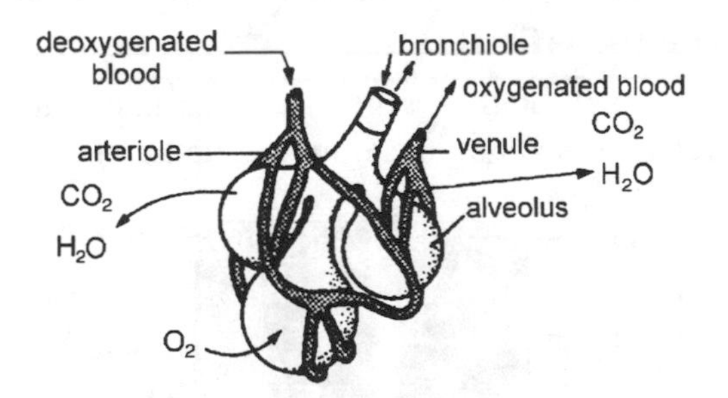

Figure 7.2 Representation of alveoli where carbon dioxide and water are eliminated

C) Liver – The liver is the gland that secretes bile for the emulsification of fats. It is considered an organ of excretion because:

1) it removes old red blood cells from the bloodstream.
2) amino acids in excess of the body's anabolic needs are deaminated by the liver.
3) all monosaccharides, except glucose, are removed by the liver.

D) Urinary System – The urinary system of humans consists of a pair of kidneys, a pair of ureters, the urinary bladder, and the urethra.

1) Kidneys – The kidneys are located against the dorsal body wall just below the diaphragm. They are composed of three distinct regions: the cortex, medulla, and pelvis. The capillaries and tubules in the kidneys form nephrons which remove metabolic wastes from the blood. Blood reaches the kidneys via a right and left renal artery and leaves via right and left renal veins.

a) nephron – structural and functional unit of the kidney which manufactures urine

b) glomerulus – network of capillaries which constitutes part of a single nephron

c) Bowman's capsule – double-walled chamber which surrounds the glomerulus

d) proximal tubule – segment of the nephron tubule which reclaims sodium ions, glucose, and amino acids

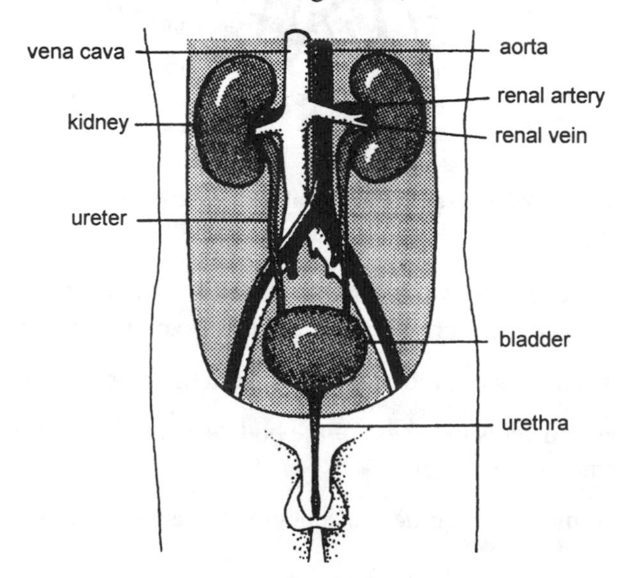

Figure 7.3 The human excretory system

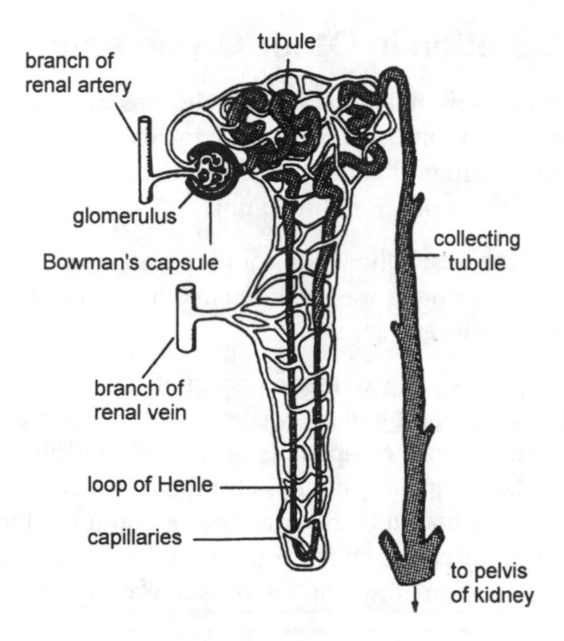

Figure 7.4 A single nephron

e) loop of Henle – The proximal tubule leads into the loop of Henle where sodium ions are actively transported out of the segment.

f) distal tubule – Additional sodium can be pumped out by the distal tubule because it is variably permeable to water.

g) collecting tubule – receives urine from smaller tubules

2) Ureters – Urine flows down from the kidney to the bladder by means of the ureter.

3) Bladder – The bladder is a hollow, muscular organ which is capable of expanding when urine flows into it.

4) Urethra – Urine flows to the outside from the bladder by way of the urethra.

7.3 Excretion in Other Organisms

A) Protozoans – Elimination of metabolic wastes occurs by diffusion through the plasma membrane. The major waste products include ammonia, carbon dioxide, mineral salts, and water. Some have contractile vacuoles for elimination.

B) Hydra – The metabolic wastes of the hydra are excreted by simple diffusion. The major wastes of the hydra include ammonia, water, carbon dioxide, and salts.

C) Grasshopper – The excretory system of the grasshopper is made up of Malpighian tubules. Wastes such as water, salts, and dissolved nitrogenous compounds diffuse into the blood in the body cavity. The Malpighian tubules absorb these wastes. Water present in the Malpighian tubules is reabsorbed into the blood. Any remaining waste passes into the intestine where it is eliminated. Salts, uric acid, and small quantities of water are the major metabolic wastes.

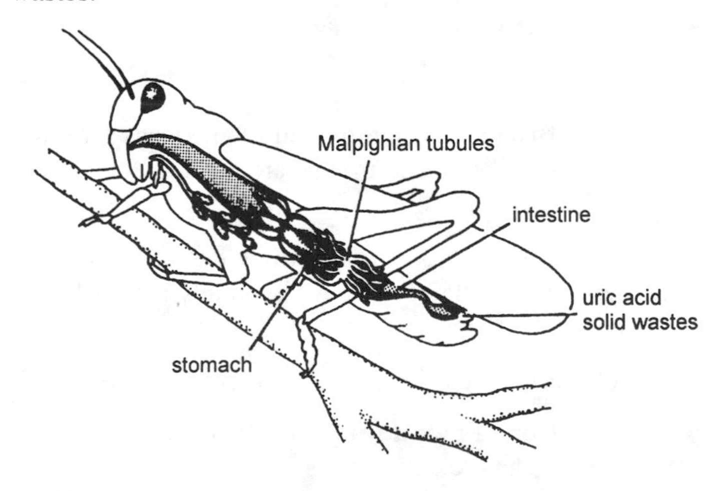

Figure 7.5 Excretory organs of a grasshopper

D) Earthworm – The nephridia are the major excretory structures of the earthworm. They are needed for the elimination of water, urea, ammonia, and mineral salts. Each nephridium consists of a nephrostome which lies within the body cavity. This body cavity is filled with fluid which enters the nephrostome and passes down a long tubule. As it travels down this tubule, useful materials are reclaimed by cells which line the tubule. Useless materials leave the earthworm by the nephridiopores, which are openings to the outside.

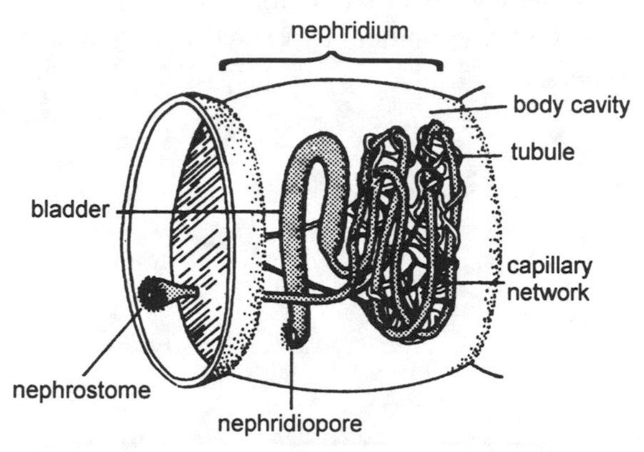

Figure 7.6 Excretory system of the earthworm

Table 7.1 Summary of Excretion

Organism	Major Organs for Excretion	Major Waste Products
Protozoans	Plasma membrane	water, ammonia, carbon dioxide, salts
Hydra	Plasma membrane	water, ammonia, carbon dioxide, salts
Grasshopper	Malpighian tubules and intestine	uric acid, salts, very little water
Earthworm	Nephridia and intestine	urea, water, salts, ammonia
Human	Lungs, kidneys, skin, liver, large intestine	urea, water, bile salts, carbon dioxide

7.4 Homeostasis

Homeostasis is the automatic maintenance of a steady state within the bodies of all organisms. It is the tendency of organisms to maintain constant the conditions of their internal environment, by responding to both internal and external changes. The kidney, for example, maintains a constant environment by excreting certain substances and conserving others.

CHAPTER 8

Hormonal Control in Animals and Plants

8.1 Hormones

A hormone is a chemical substance secreted by specific cells in one area of the body to be used in another area known as the target organ. The target organ responds by being either stimulated or inhibited.

Endocrine glands, or ductless glands, secrete their products directly into the capillaries and become part of the bloodstream.

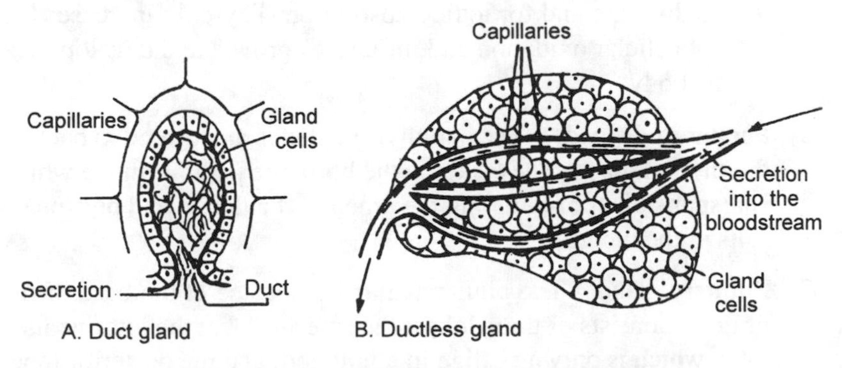

Figure 8.1 Exocrine and endocrine glands

Exocrine glands secrete hormones which pass through ducts, to reach the place where they function. They are also called duct glands.

8.2 The Mechanism and Action of Hormones

A) Only small concentrations are needed for a hormone to be effective.

B) The hormones secreted by an endocrine gland are influenced by certain substances present in the blood and by nerve impulses from the autonomic nervous system.

C) The functioning of one endocrine gland will also affect the functioning of other glands in the body.

D) Current theories include the major principle that hormones combine with specific receptors when the hormone enters the cell.

8.3 The Human Endocrine System

The major glands of the human endocrine system include the thyroid gland, parathyroid glands, pituitary gland, pancreas, adrenal glands, pineal gland, thymus gland, and the sex glands.

A) Thyroid gland – The thyroid gland is a two-lobed structure located in the neck. It is responsible for the secretion of the hormone thyroxin and for iodide absorption. Thyroxin increases the rate of cellular oxidation and influences growth and development of the body.

B) Parathyroid glands – The parathyroid glands are located in back of the thyroid gland. They secrete the hormone parathormone which is responsible for regulating the amount of calcium and phosphate salts in the blood.

C) Pituitary gland – The pituitary gland is located at the base of the brain. It consists of three lobes: the anterior lobe; the intermediate lobe, which is only a vestige in adulthood; and the posterior lobe.

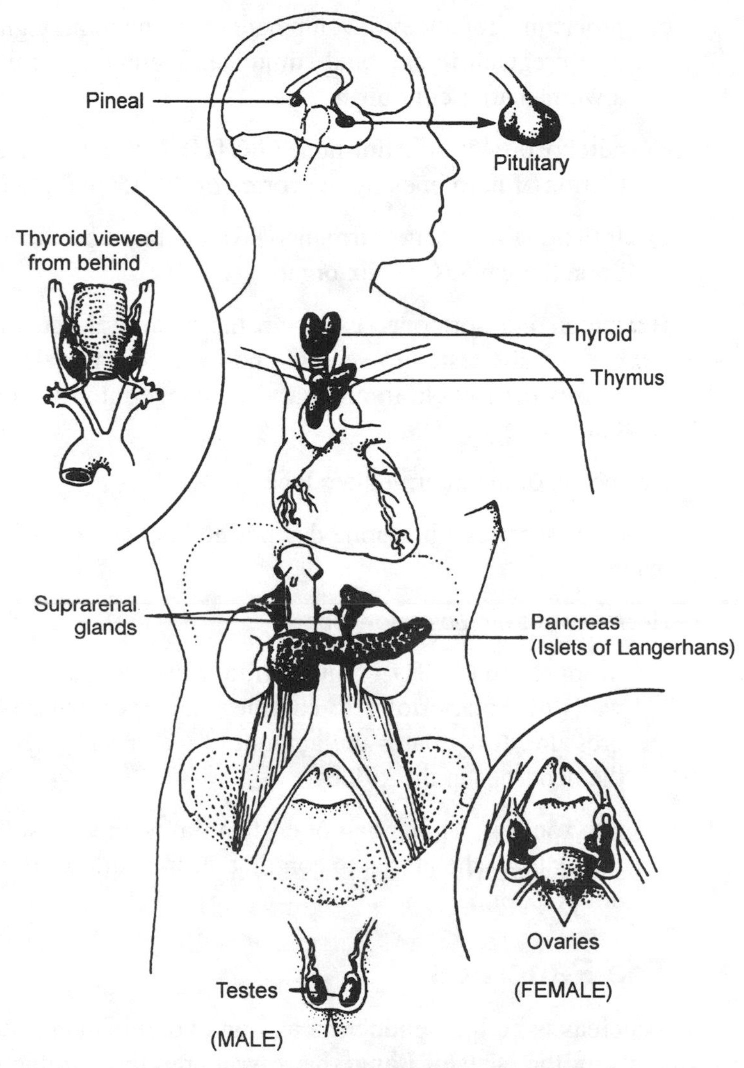

Figure 8.2 The human endocrine system

1) Hormones of the anterior lobe

   a) growth hormone – stimulates growth of bones

   b) thyroid-stimulating hormone (TSH) – stimulates the thyroid gland to produce thyroxin

c) prolactin – regulates development of the mammary glands of a pregnant female and stimulates secretion of milk in a woman after childbirth

d) adrenocorticotropic hormone (ACTH) – stimulates the secretion of hormones by the cortex of the adrenal glands

e) follicle-stimulating hormone (FSH) – this hormone acts upon the gonads, or sex organs

f) luteinizing hormone (LH) – in the male, LH causes the cells in the testes to secrete androgens. In females, LH causes the follicle in an ovary to change into the corpus luteum.

2) Hormones of the intermediate lobe

This lobe secretes a hormone that has no known effects in humans.

3) Hormones of the posterior lobe

a) Vasopressin (ADH) – This hormone causes the muscular walls of the arterioles to contract, thus increasing blood pressure. It regulates the amount of water reabsorbed by the nephrons in the kidney.

b) Oxytocin – This hormone stimulates the muscle of the walls of the uterus to contract during childbirth. It induces labor.

8.4 The Pancreas

The pancreas is both an endocrine and an exocrine gland. As an endocrine gland, the Islets of Langerhans, scattered through the pancreas, secrete insulin and glucagon.

A) Insulin – acts to lower the level of glucose in the bloodstream. Glucose is converted to glycogen.

B) Glucagon – increases the level of glucose in the blood by helping to change liver glycogen into glucose.

8.5 Adrenal Glands

The two adrenal glands are located on top of each kidney. They are composed of two regions: the adrenal cortex and the adrenal medulla.

A) Hormones of the Adrenal Cortex

1) cortisones – regulate the change of amino acids and fatty acids into glucose. They also help to suppress reactions that lead to the inflammation of injured parts.

2) cortins – regulate the use of sodium and calcium salts by the body cells.

3) sex hormones – they are similar in chemical composition to hormones secreted by sex glands.

B) Hormones of the Adrenal Medulla

1) epinephrine – this hormone is responsible for the release of glucose from the liver, the relaxation of the smooth muscles of the bronchioles, dilation of the pupils of the eye, a reduction in the clotting time of blood, and an increase in the heart beat rate, blood pressure and respiration rate.

2) norepinephrine – this hormone is responsible for the constriction of blood vessels.

8.6 Pineal Gland

The pineal gland is attached to the brain above the cerebellum. It is responsible for the production of melatonin whose role in humans is uncertain.

8.7 Thymus Gland

The thymus gland is located under the breastbone. Although there is no convincing evidence for its role in the human adult, it does secrete thymosin hormone in infants which stimulates the formation of an antibody system.

8.8 Sex Glands

These glands include the testes of the male and the ovaries of the female.

A) Testes – Luteinizing hormone stimulates specific cells of the testes to secrete androgens. Testosterone, which controls the development of male secondary sex characteristics, is the principle androgen.

B) Ovaries – Estrogen is secreted from the cells which line the ovarian follicle. This hormone is responsible for the development of female secondary sex characteristics.

Table 8.1
Human Endocrine Glands and their Functions

Gland	Hormone	Function
Pituitary	Growth hormone	Stimulates growth of skeleton
Anterior lobe	FSH	Stimulates follicle formation in ovaries and sperm formation in testes
	LH	Stimulates formation of corpus luteum in ovaries and secretion of testosterone in testes
	TSH	Stimulates secretion of thyroxin from thyroid gland
	ACTH	Stimulates secretion of cortisone and cortin from adrenal cortex
	Prolactin	Stimulates secretion of milk in mammary glands
Posterior lobe	Vasopressin (ADH)	Controls narrowing of arteries and rate of water absorption in kidney tubules
	Oxytocin	Stimulates contraction of smooth muscle of uterus

Table 8.1 (continued)

Thyroid	Thyroxine	Controls rate of metabolism and physical and mental development
	Calcitonin	Controls calcium metabolism
Parathyroids	Parathormone	Regulates calcium and phosphate level of blood
Islets of Langerhans		
Beta cells	Insulin	Promotes storage and oxidation of glucose
Alpha cells	Glucagon	Releases glucose into bloodstream
Thymus	Thymus hormone	Stimulates formation of antibody system
Adrenals Cortex	Cortisones	Promote glucose formation from amino acids and fatty acids
	Cortins	Control water and salt balance
	Sex hormones	Influence sexual development
Medulla	Epinephrine (adrenalin) or norepinephrine (noradrenalin)	Releases glucose into bloodstream, increases rate of heartbeat, increases rate of respiration, reduces clotting time, relaxes smooth muscle in air passages
Gonads Ovaries, follicle cells	Estrogen	Controls female secondary sex characteristics
Corpus luteum cells	Progesterone	Helps maintain attachment of embryo to mother
Testes	Testosterone	Controls male secondary sex characteristics

8.9 Endocrine Abnormalities

A) Myxedema – This results when there is a deficiency in the amount of thyroxine that is secreted by the thyroid gland. Myxedema occurs in adults and is characterized by decreased heat production and a low metabolic rate.

B) Cretinism – Cretinism occurs when the thyroid gland is defective at birth. It is a type of dwarfism characterized by defective teeth, protruding abdomen, and low mental ability.

C) Goiter – An enlarged thyroid gland is known as a goiter. Nontoxic goiter is the result of a lack of iodine in the diet. Toxic goiter results from an overdevelopment of the thyroid gland.

D) Hyperparathyroidism – An oversecretion of the parathyroid hormone is responsible for excessive withdrawal of calcium from the bones, causing them to soften and break easily.

E) Hypoparathyroidism – Removal of the parathyroids results in a high phosphate concentration and a low calcium concentration in the blood. This produces serious disturbances of muscles and nerves.

F) Diabetes – Diabetes results when there is not a sufficient amount of insulin secreted by the beta cells. The insulin cannot regulate the passage of glucose from the blood into the muscles and the liver.

G) Giantism – Giantism results when there is an oversecretion of growth hormone during childhood.

H) Dwarfism – Dwarfism results when there is an undersecretion of growth hormone during childhood.

I) Acromegaly – Acromegaly is the result of an oversecretion of growth hormone in the adult. This results in an overgrowth of parts of the body which can still respond to the hormone such as the ends of the bones of the hands, feet, and face.

8.10 Plant Hormones

A) Auxins – These plant growth regulators stimulate the elongation of specific plant cells and inhibit the growth of other plant cells.

B) Gibberellins – In some plants, gibberellins are involved in the stimulation of flower formation. They also increase the stem length of some plant species and the size of fruits. Gibberellins also stimulate the germination of seeds.

C) Cytokinins – Cytokinins increase the rate of cell division and stimulate the growth of cells in a tissue culture. They also influence the shedding of leaves and fruits, seed germination, and the pattern of branch growth.

CHAPTER 8–A

Relevant Problems and their Solutions

The following problems are typical of questions you will encounter in an introductory biology course. Following each problem is a detailed solution.

- **PROBLEM 1:**

  Electrons are located at definite energy levels in orbitals around the nucleus of an atom. Sometimes electrons move from one orbital to another. What happens to cause such a change?

Solution: When an electron is in its ground state in the atom, it is occupying the orbital of lowest energy for that electron. There may be other electrons in the atom that are located in orbitals of even lower energy. The further the electron is from the nucleus, the greater the energy associated with that electron. An electron in the ground state may absorb energy from an external source (e.g., heat), and be promoted to a higher-energy orbital. The electron would now be considered to be in an excited state. In a similar fashion, an excited electron may drop down to a lower-energy orbital (not necessarily the ground state), and in the process will decrease its energy content. This energy difference will be released in the form of a photon of light.

• **PROBLEM 2:**

What is meant by cellular metabolism? How does metabolism differ from anabolism and catabolism?

Solution: Cellular metabolism includes the following processes that transform substances extracted from the environment: degradation, energy production, and biosynthesis. All heterotrophic organisms break down materials taken from their environment, and utilize the product to synthesize new macromolecules. When materials are broken down, energy is released or stored in the cell; when the products are used in syntheses, energy is expended.

There are two general types of metabolism. That part of metabolism by which new macromolecules are synthesized with the consumption of energy is termed anabolism (In Greek, ana = up, as in build up). The degradation reactions, which decompose ingested material and release energy, are collectively termed catabolism (In Greek, cata = down, as in break down). The degradation of a glucose molecule to carbon dioxide and water during aerobic respiration is an example of catabolism. In the process, 38 molecules of ATP are formed for later use if needed. The degradation of fats is also an example of catabolism. The biosynthesis of proteins (from amino acids) and of carbohydrates like starch or glycogen, (from simple sugars) are two important anabolic processes.

• **PROBLEM 3:**

What are some of the important properties and characteristics of enzymes?

Solution: An important property of enzymes is their catalytic ability. Enzymes control the speed of many chemical reactions that occur in the cell. To understand the efficiency of an enzyme, one can measure the rate at which an enzyme operates—also called the turnover number. The turnover number is the number of molecules of substrate which is acted upon by a molecule of enzyme per second. Most enzymes have high turnover numbers and are thus needed in the cell in relatively small amounts. The maximum turnover number of catalase, an enzyme which decomposes hydrogen peroxide, is 10^7 molecules/sec. It would require years for an iron atom to accomplish the same task.

A second important property of enzymes is their specificity, that is, the number of different substrates they are able to act upon. The surface of the enzyme reflects this specificity. Each enzyme has a region called a binding site to which only certain substrate molecules can bind efficiently. There are varying degrees of specificity: urease, which decomposes urea to ammonia and carbon dioxide, will react with no other substance; however lipase will hydrolyze the ester bonds of a wide variety of fats.

Another aspect of enzymatic activity is the coupling of a spontaneous reaction with a non-spontaneous reaction. An energy-requiring reaction proceeds with an increase in free energy and is non-spontaneous. To drive this reaction, a spontaneous energy-yielding reaction occurs at the same time. The enzyme acts by harnessing the energy of the energy-yielding reaction and transferring it to the energy-requiring reaction.

The structure of different enzymes differs significantly. Some are composed solely of protein (for example, pepsin). Others consist of two parts: a protein part (also called an apoenzyme) and a non-protein part, either an organic coenzyme or an inorganic cofactor, such as a metal ion. Only when both parts are combined can activity occur.

There are other important considerations. Enzymes, as catalysts, do not determine the direction a reaction will go, but only the rate at which the reaction reaches equilibrium. Enzymes are efficient because they are needed in very little amounts and can be used repeatedly. As enzymes are proteins, they can be permanently inactivated or denatured by extremes in temperature and pH, and also have an optimal temperature or pH range within which they work most efficiently.

- **PROBLEM 4:**

What roles does glucose play in cell metabolism?

Solution: Glucose, a six carbon monosaccharide, is the primary source of energy for all cells in the body. The complete oxidation of one molecule of glucose yields 36 ATP molecules in a cell. The energy stored in the form of ATP can then be used in a variety of ways, including the contraction of a muscle or the secretion of an enzyme. The supply of glucose to all cells must be maintained at certain minimal levels so that every cell gets an adequate amount of glucose. Brain cells, unable to store glucose, are the first to suffer when the blood glucose concentration falls below a certain critical level. On the other hand, muscle cells are less affected by changes in the glucose level of the blood because of their local storage of glucose as glycogen.

In certain cells, glucose can be converted to glycogen and can be stored only in this form, because glucose will diffuse into and out of a cell readily. Glycogen is a highly branched polysaccharide of high molecular weight, composed of glucose units linked by α-glycosidic bonds. Once the conversion is complete, glycogen remains inside the cell as a storage substance. When glucose is needed, glycogen is reconverted to glucose.

In addition to being stored as glycogen or oxidized to generate energy, glucose can be converted into fat for storage. For example, after a heavy meal causes the supply of glucose to exceed the immediate need, glucose molecules can be transformed into fat which is stored in the liver and/or adipose (fat) tissue. This fat serves as an energy supply for later use.

• PROBLEM 5:

Describe the path of a molecule of sugar from the time it enters the mouth as part of a molecule of starch, until it reaches the cytoplasm of the cells lining the small intestine.

Solution: Upon entering the mouth, the starch is chewed and mixed with saliva. The saliva dissolves small molecules and coats the larger clumps of starch with a solution of the enzyme amylase. As the mass of food is swallowed, the enzyme amylase from the saliva hydrolyzes the starch, degrading it into maltose units. Maltose is a disaccharide composed of two glucose molecules. Eventually, the acidity of the stomach inhibits the actions of the salivary amylase, leaving most of the starch undigested. No further digestion of the starch occurs until it reaches the small intestine. There, pancreatic amylase finishes the cleavage of starch into maltose. Intestinal maltase then hydrolyzes the maltose molecules into two glucose molecules each. Other intestinal enzymes cleave other disaccharides, such as sucrose, into monosaccharides.

When these free sugar molecules come into contact with the cells lining the small intestine, they are held by binding enzymes on the cell surface. Glucose is transported through the membrane into the cellular cytoplasm.

• PROBLEM 6:

What are the functions of roots? What are the two types of root systems?

Solution: Roots serve two important functions: one is to anchor the plant in the soil and hold it in an upright position; the second and biologically more important function is to absorb water and minerals from the soil and conduct them to the stem. To perform these two functions, roots branch and rebranch extensively through the soil resulting in an enormous total surface area which usually exceeds that of the stem's. Roots can be classified as a taproot system (i.e. carrots, beets) in which the primary (first) root increases in diameter and length and functions as a storage place for large quantities of food. A fibrous root system is composed of many thin main roots of equal size with smaller branches.

Additional roots that grow from the stem or leaf, or any structure other than the primary root or one of its branches are termed adventitious roots. Adventitious roots of climbing plants such as the ivy and other vines attach the plant body to a wall or a tree. Adventitious roots will arise from the stems of many plants when the main root system is removed. This accounts for the ease of vegetative propagation of plants that are able to produce adventitious roots.

• PROBLEM 7:

Describe feeding in planarians.

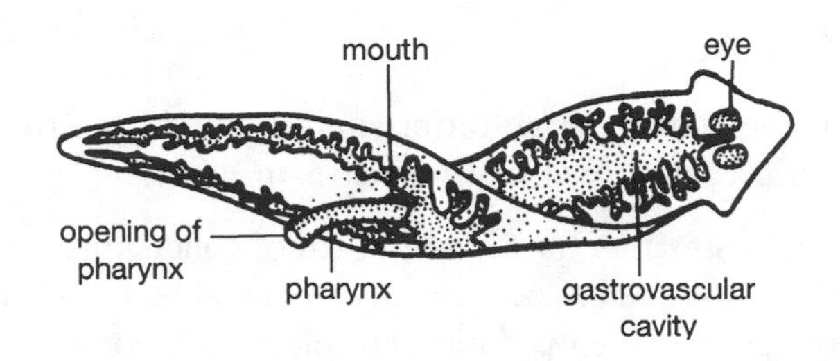

Planaria, showing much-branched gastrovascular cavity and extruded pharynx.

Solution: Planarians belong to the group of fresh water flatworms and have a digestive system which lacks an anal opening. The mouth of the planarian opens into a cavity that contains the muscular pharynx, or proboscis, which can be protruded through the mouth directly onto the prey (see figure). In feeding, the planarian moves over its food object, which may be a small worm, crustacean, or insect larva, and traps it with its body. The pharynx is then extended and attached to the food material, and by sucking movements produced by muscles in the pharynx, the food is torn into bits and ingested. Digestive enzymes are released in order to assist the planarian in breaking down the food prior to ingestion. The pharynx delivers the food into the three-branched gastrovascular cavity. The branching of the planaria's gastrovascular cavity into one anterior and two posterior branches provides for the distribution of the end products of digestion to all parts of the body. Flatworms with three branched gut cavities are called triclads, in contrast to those with many branches, called polyclads.

Most of the digestion in planaria is intracellulary which means that it occurs in food vacuoles in cells lining the digestive cavity. The end products of digestion diffuse from these cells throughout the tissues of the body. Undigested materials are eliminated by the planaria through its mouth. The mouth, then serves as both the point of ingestion and the point of egestion.

It is interesting to note that planarians can survive without food for months, gradually digesting their own tissues, and growing smaller as time passes.

- **PROBLEM 8:**

Why is it that alveolar air differs in its composition from atmospheric air? Of what significance is this fact?

Solution: The respiratory tract is composed of conducting airways and the alveoli. Gas exchange occurs only in the alveoli and not in the conducting airways. The maximum alveolar volume is about 500 ml. Let us consider then what takes place during expiration. Five hundred milliliters of air is forced out of the alveoli, and into the conducting airways of the respiratory tract. Of this air, 150 ml remains in the respiratory airways following expiration, while 350 ml of air is exhaled from the body. During the next inspiration 500 ml of air are taken up by the alveoli, but 150 ml of this air is not atmospheric, but rather is the air that remained in the tubes following the previous expiration. One can see, then, that only 350 ml of fresh atmospheric air enters the alveoli during each inspiration. At the end of inspiration, 150 ml of fresh air also fills the airways but does not reach the alveoli. Hence no gas exchange between this air and the blood can occur. This fresh air will be expelled from the body during the next expiration, and will be replaced again with 150 ml of alveolar air, thus completing the cycle. From this cycle it can be seen that of the 500 ml of air entering the body during inspiration, 150 ml of it never reaches the alveoli, but instead remains in the conducting tubes of the respiratory system. The term "anatomical dead space" is given to the space within the conducting tubes because no gas exchange with the blood can take place there.

The question arises then, as to the significance of this dead space. Because the tubes are not completely emptied and filled with each breath, "new" air can mix with "old" air. Consequently, alveolar air contains less oxygen and more carbon dioxide than atmospheric air. In addition, the air that remains in the alveoli following expiration (the residual volume) can, to a limited extent, mix with the incoming air and thereby alter its composition.

• PROBLEM 9:

In aquatic animals, external respiration is carried out by specialized structures called gills. How do these gills operate?

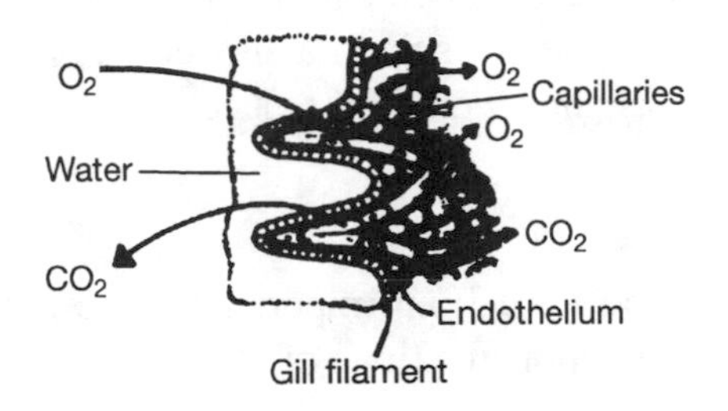

Respiration in fish

<u>Solution</u>: Fish, molluscs, and many arthropods (except insects) possess gills for respiratory purposes (see figure). Organisms such as these require a mechanism to keep a fresh supply of water flowing about them. This is necessary because gas exchange occurs between the blood vessels in the gills and the water which contains dissolved oxygen. Continuous bathing of the gills ensures that there will always be enough oxygen available for the organism. It also ensures that the carbon dioxide diffusing from the gills into the water is removed from the organism. This is important in that the CO_2 concentration gradient is maintained, so diffusion of CO_2 will continue to occur from the gills to the water.

A fish opens its mouth and receives a quantity of water. It then closes its mouth and forces the water out past its gills by contracting its oral cavity. Gills have thin walls, are moist, and are well supplied with blood capillaries. Oxygen, which is dissolved in the water, diffuses though the gill epithelium and into the blood capillaries. Simultaneously, carbon dioxide diffuses in the opposite direction in accordance with the CO_2 concentration gradient. The amount of oxygen dissolved in sea water is relatively constant, but the amount in freshwater ponds may fluctuate greatly. A fish would suffocate in water that is lacking sufficient amounts of dissolved oxygen.

It is interesting to note that the direction of blood flow through a gill is opposite to the direction of water flow over the gill. This counter-current system maximizes the amount of gas exchange that can take place. If both the blood flow and water current were in the same direction, the CO_2 and O_2 concentration gradients across the gills would decrease. This would result in a slower rate of diffusion, and the amount of gas exchange would be reduced.

- **PROBLEM 10:**

Explain why blood is such an important tissue in many animals. Discuss the major functions of blood.

Solution: All cells, in order to survive, must obtain the necessary raw materials for metabolism, and have a means for the removal of waste products. In small plants and animals living in an aquatic environment, these needs are provided for by simple diffusion. The cells of such organisms are very near the external watery medium, and so nutrients and wastes do not have a large distance to travel. However, as the size of the organism increases, more and more cells become further removed from the media bathing the peripheral cells. Diffusion cannot provide sufficient means for transport. In the absence of a specialized transport system, the limit on the size of an aerobic organism would be about a millimeter, since the diffusion of oxygen and nutrients over great distances would be too slow to meet the metabolic needs of all the cells of the organism. In addition, without internal transport, organisms are restricted to watery environments, since the movement to land requires an efficient system for material exchange in non-aqueous surroundings. Therefore, larger animals have developed a system of internal transport, the circulatory system. This system, consisting of an extensive network of various vessels, provides each cell with an opportunity to exchange materials by diffusion.

Blood is the vital tissue in the circulatory system, transporting nutrients and oxygen to all the cells and removing carbon dioxide and other wastes from them. Blood also serves other important functions. It transports hormones, the secretions of the endocrine glands, which affect organs sensitive to them. Blood also acts to regulate the acidity and alkalinity of the cells via control of their salt and water content. In addition, the blood acts to regulate the body temperature by cooling certain organs and tissues when an excess of heat is produced (such as in exercising muscle) and warming tissues where heat loss is great (such as in the skin).

Some components of the blood act as a defense against bacteria, viruses, and other pathogenic (disease-causing) organisms. The blood also has a self-preservation system, called a clotting mechanism, so that loss of blood due to vessel rupture is reduced.

• PROBLEM 11:

Why does the number of red blood cells in the human body increase at high altitudes?

Solution: It has been observed that the loss of red cells by hemorrhage decreases the ability of the blood to deliver sufficient oxygen, and increased erythrocyte production results. Also, partial destruction of the bone marrow (by X-ray, for example) destroys the sites of erythrocyte production and diminishes the availability of oxygen to the tissues. Increased red cell production by the remaining healthy marrow follows. It seems that the initiator or stimulus for increased erythrocyte production could be a lack of oxygen. This is indeed found to be true. At high altitudes, the partial pressure of oxygen is decreased, and thus less oxygen is delivered to the tissues. Any condition that causes a decrease in the amount of oxygen transported to the tissues causes an increased rate of erythrocyte production.

Actually, a decreased oxygen concentration in the bone marrow does not directly (by itself) expedite erythrocyte production. In response to decreased oxygen levels, the kidney, liver, and other tissues secrete erythropoietin, also called the erythropoietic stimulating factor. This glycoprotein is transported via the blood to the bone marrow, where it stimulates erythrocyte production. When the number of red blood cells has increased to the point where the oxygen level in the tissues is again normal, erythropoietin secretion decreases and a normal amount of erythrocytes are again produced. This negative feedback mechanism prevents an overproduction of red cells, which keeps the viscosity of the blood from increasing to a dangerous level.

It is interesting to note that erythrocyte production and hemoglobin synthesis are not necessarily correlated. An iron deficiency decreases hemoglobin synthesis (since the heme part of the hemoglobin molecule contains an iron atom). This low hemoglobin level leads to a reduction of oxygen brought to the tissues. As was cited before, a reduced oxygen level causes increased red blood cell production. In this case, the red blood cells have a lower hemoglobin content than normal (they are called hypochromic erythrocytes) and thus are less efficient oxygen carriers.

• PROBLEM 12:

The circulatory system of insects does not function in gas exchange. What is its function? Describe the circulatory and respiratory systems in insects.

Solution: Insects, which have high metabolic rates, need oxygen in large amounts. However, insects do not rely on the blood to supply oxygen to their tissues. This function is fulfilled by the tracheal system. The blood serves only to deliver nutrients and remove wastes.

The insect heart is a muscular dorsal tube, usually located within the first nine abdominal segments. The heart lies within a pericardial sinus. The pericardial sinus is not derived from the coelom, but is instead a part of the hemocoel. It is separated by connective tissue from the perivisceral sinus which is the hemocoel surrounding the other internal structures. Usually, the only vessel besides the heart is an anterior aorta. Blood flow is normally posterior to anterior in the heart and anterior to posterior in the perivisceral sinus. Blood from the perivisceral sinus drains into the pericardial sinus. The heart is pierced by a series of openings or ostia, which are regulated by valves, so that blood only flows in one direction. When the heart contracts, the ostia close and blood is pumped forward. When the heart relaxes, the ostia open and blood from the pericardial sinus is drawn into the heart through the ostia. After leaving the heart and aorta, the blood fills the spaces between the internal organs, bathing them directly. The rate of blood flow is regulated by the motion of the muscles of the body wall or the gut.

A respiratory system delivers oxygen directly to the tissues in the insect. A pair of openings called spiracles is present on the first seven or eight abdominal segments and on the last one or two thoracic segments. Usually, the spiracle is provided with a valve for closing and with a filtering apparatus (composed of bristles) to prevent entrance of dust and parasites.

The organization of the internal tracheal system is quite variable, but usually a pair of longitudinal trunks with cross connections is found. Larger tracheae are supported by thickened rings of cuticle, called taenidia. The tracheae are widened in various places to form internal air sacs. The air sacs have no taenidia and are sensitive to ventilation pressures (see below). The tracheae branch to form smaller and smaller subdivisions, the smallest being the tracheoles. The smallest tracheoles are in direct contact with the tissues and are filled with fluid at their tips. This is where gas exchange takes place.

Within the tracheae, gas transport is brought about by diffusion, ventilation pressures, or both. Ventilation pressure gradients result from body movements. Body movements causing compression of the air sacs and certain elastic tracheae force air out; those causing expansion of the body wall result in air rushing into the tracheal system. In some insects, the opening and closing of spiracles is coordinated with body movements. Grasshoppers, for example, draw air into the first four pairs of spiracles as the abdomen expands, and expel air through the last six pairs of spiracles as the abdomen contracts.

• PROBLEM 13:

The terms *defecation*, *excretion*, and *secretion* are sometimes confused. What is meant by these terms?

Solution: Defecation refers to the elimination of wastes and undigested food, collectively called feces, from the anus. Undigested food materials have never entered any of the body cells and have not taken part in cellular metabolism; hence they are not metabolic wastes. Excretion refers to the removal of metabolic wastes from the cells and bloodstream. The excretion of wastes by the kidneys involves an expenditure of energy by the cells of the kidney, but the act of defecation requires no such effort by the cells lining the large intestine. Secretion refers to the release from a cell of some substance which is utilized either locally or elsewhere in some body processes; for example, the salivary glands secrete saliva, which is used in the mouth in the first step of chemical digestion. Secretion also involves cellular activity and requires the expenditure of energy by the secreting cell.

• PROBLEM 14:

Most marine vertebrates have body fluids with osmotic pressure lower than that of their saline environment. How do these organisms osmoregulate in face of a perpetual threat of dehydration?

Solution: To survive their hostile saline environment, marine organisms show a variety of adaptations. The concentration of body fluids in marine fish is about one-third that of the sea water. Osmotic pressure, being higher in sea water, tends to drive water out of the body fluid of the fish. One site of severe water loss is the gill, which is exposed directly to the surrounding water for gaseous diffusion. To compensate for this loss, the fish must take in large amounts of water. However, the only source of water available is sea water; if consumed, the sea water would cause further water loss from the body cells because of its higher osmotic pressure. Marine fish, however, overcome this problem. In the gills, excess salt from the consumed sea water is actively transported out of the blood and passed back into the sea.

Marine birds and reptiles also face the same problem of losing water. Sea gulls and penguins take in much sea water along with the fish which they scoop from the sea. To remove the excess salt, there are glands in the bird's head which can secrete saltwater having double the osmolarity of sea water. Ducts from these glands lead into the nasal cavity and the saltwater drips out from the tip of the bill. The giant sea turtle, a marine reptile, has similar glands but the ducts open near the eyes. Tears from the eyes of turtles are not an emotional response, in contrast to popular belief. Rather, "crying" is a physiological mechanism to get rid of excess salt.

• PROBLEM 15:

What hormones are involved with the changes that occur in a pubescent female and male?

Solution: Puberty begins in the female when the hypothalamus stimulates the anterior pituitary to release increased amounts of FSH (follicle stimulating hormone) and LH (luteinizing hormone). These hormones cause the ovaries to mature, which then begin secreting estrogen and progesterone, the female sex hormones. These hormones, particularly estrogen, are responsible for the development of the female secondary sexual characteristics. These characteristics include the growth of pubic hair, an increase in the size of the uterus and vagina, a broadening of the hips and development of the breasts, a change in voice quality, and the onset of the menstrual cycles.

Before the onset of puberty in the male, no sperm and very few male sex hormones are produced by the testes. The onset of puberty begins, as in the female, when the hypothalamus stimulates the anterior pituitary to release increased amounts of FSH and LH. In the male, FSH stimulates maturation of the seminiferous tubules which produce the sperm. LH is responsible for the maturation of the interstitial cells of the testes. It also induces them to begin secretion of testosterone, the male sex hormone. When enough testosterone accumulates, it brings about the whole spectrum of secondary sexual characteristics normally associated with puberty. These include growth of facial and pubic hair, deepening of the voice, maturation of the seminal vesicles and the prostate gland, broadening of the shoulders, and the development of the muscles.

If the testes were removed before puberty, the secondary sexual characteristics would fail to develop. If they were removed after puberty, there would be some retrogression of the adult sexual characteristics, but they would not disappear entirely.

• PROBLEM 16:

What is a pheromone, and how does it differ from a hormone?

Solution: The behavior of animals may be influenced by hormones, organic chemicals that are released into the internal environment by endocrine glands, and which regulate the activities of other tissues located some distances away. Animal behavior is also controlled by pheromones—substances that are secreted by exocrine glands into the external environment. Pheromones influence the behavior of other members of the same species. Pheromones represent a means of communication and of transferring information by smell or taste. Pheromones evoke specific behavioral, developmental, or reproductive responses in the recipient; these responses may be of great significance for the survival of the species.

Pheromones act in a specific manner upon the recipient's central nervous system, and produce either a temporary or a long-term effect on its development or behavior. Pheromones are of two classes: releaser pheromones and primer pheromones. Among the releaser pheromones are the sex attractants of moths and the trail pheromones secreted by ants, which may cause an immediate behavioral change in conspecific individuals. Primer pheromones act more slowly and play a role in the organism's growth and differentiation. For example, the growth of locusts and the number of reproductive members and soldiers in termite colonies are all controlled by primer pheromones.

• PROBLEM 17:

What are the difficulties in the classification of tissues of higher plants?

Solution: The characteristics of plant cells themselves make it difficult for botanists to agree on any one classification system. The different types of cells intergrade and a given cell can change from one type to another during the course of its life. As a result, the tissues formed from such cells also intergrade and can share functional and structural characteristics. Some plant tissues contain cells of one type while others consist of a variety of cell types. Plant tissues cannot be fully characterized or distinguished on the basis of any one single factor such as location, function, structure, or evolutionary heritage. Plant tissues can be divided into two major categories: meristematic tissues, which are composed of immature cells and are regions of active cell division; and permanent tissues, which are composed of more mature, differentiated cells. Permanent tissues can be subdivided into three classes of tissues—surface, fundamental, and vascular. However, the classification of plant tissues into categories based purely on their maturity runs into some difficulties. Some permanent tissues may change to meristematic activity under certain conditions. Therefore, this classification is not absolutely reliable.

• PROBLEM 18:

What is the structure and function of the chloroplasts in green plants?

Solution: The chloroplasts have the ability to transform the energy of the sun into chemical energy stored in bonds that hold together the atom of such foods (fuel) as glucose. By the process of photosynthesis, carbon dioxide and water react to produce carbohydrates with the simultaneous release of oxygen. Photosynthesis, which occurs in the chloroplasts, is driven forward by energy obtained from the sun.

Photosynthesis involves two major sets of reactions, each consisting of many steps. One set depends on light and cannot occur in the dark; hence, this set is known as the light reaction. Photophosphorylation is responsible for the production of ATP from sunlight and the release of oxygen derived from water. The other set, referred to as the dark reaction, is not dependent on light. In the dark reaction, carbon dioxide is reduced to carbohydrates using the energy of ATP from the light reactions.

• PROBLEM 19:

Even though there is no such thing as a "typical cell"—for there are too many diverse kinds of cells—biologists have determined that there are two basic cell types. What are these two types of cells?

Solution: Cells are classified as either prokaryotic or eukaryotic. Prokaryotes are strikingly different from eukaryotes in their ultrastructural characteristics. A key difference between the two cell types is that prokaryotic cells lack the nuclear membrane characteristic of eukaryotic cells. Prokaryotic cells have a nuclear region, which consists of nucleic acids. Eukaryotic cells have a nucleus, bound by a double-layered membrane. The eukaryotic nucleus consists of DNA which is bound to proteins and organized into chromosomes.

Bacteria and blue-green algae are prokaryotic unicellular organisms. Other organisms, for example, protozoa, algae, fungi, higher plants, and animals are eukaryotic. Within eukaryotic cells are found discrete regions that are usually delimited from the rest of the cell by membranes. These are called membrane-bound subcellular organelles. They perform specific cellular functions, for example, respiration and photosynthesis. The enzymes are not segregated into discrete organelles although they have an orderly arrangement. Prokaryotic cells lack an endoplasmic reticulum, Golgi apparatus, lysosomes, and vacuoles. In short, prokaryotic cells lack the internal membranous structure characteristic of eukaryotic cells.

There are other differences between prokaryotic cells and eukaryotic cells. The ribosomes of bacteria and blue-green algae are smaller than the ribosomes of eukaryotes. The flagella of bacteria are structurally different from eukaryotic flagella. The cell wall of bacteria and blue-green algae usually contains muramic acid, a substance that plant cell walls and the cells walls of fungi do not contain.

• PROBLEM 20:

Differentiate between acids, bases, and salts. Give examples of each.

Solution: There are essentially two widely used definitions of acids and bases: the Lowry-Bronsted definition and the Lewis definition. In the Lowry-Bronsted definition, an acid is a compound with the capacity to donate a proton, and a base is a compound with the capacity to accept a proton. In the Lewis definition, an acid has the ability to accept an electron pair and a base the ability to donate an electron pair.

Salts are a group of chemical substances which generally consist of positive and negative ions arranged to maximize attractive forces and minimize repulsive forces. Salts can be either inorganic or organic. For example, sodium chloride, NaCl, is an inorganic salt which is actually best represented with its charges Na^+Cl^-; sodium acetate, CH_3COONa or $CH_3COO^-Na^+$, is an organic salt.

Some common acids important to the biological system are acetic acid (CH_3COOH), carbonic acid (H_2CO_3), phosphoric acid (H_3PO_4), and water. Amino acids, the building blocks of protein, are compounds that contain an acidic group (–COOH). Some common bases are ammonia (NH_3), pyridine (C_5H_5N), purine (NH_4), and water. The nitrogenous bases important in the structure of DNA and RNA carry the purine or pyridine functional group. Water has the ability to act both as an acid ($H_2O \rightarrow -H^+ \rightarrow OH^+$) and as a base ($H_2O + H^+ \rightarrow H_3O^+$) depending on the conditions of the reaction, and is thus said to exhibit amphiprotic behavior.

- **PROBLEM 21:**

Removal of the pituitary in young animals arrests growth owing to termination of the supply of growth hormone. What are the effects of growth hormone in the body? What is acromegaly?

Solution: The pituitary, under the influence of the hypothalamus, produces a growth-promoting hormone. One of the major effects of the growth hormone is to promote protein synthesis. It does this by increasing membrane transport of amino acids into cells, and also by stimulating RNA synthesis. These two events are essential for protein synthesis. Growth hormone also causes large increases in mitotic activity and cell division.

Growth hormone has its most profound effect on bone. It promotes the lengthening of bones by stimulating protein synthesis in the growth centers. The cartilaginous center and bony edge of the epiphyseal plates constitute growth centers in bone. Growth hormone also lengthens bones by increasing the rate of osteoblast (young bone cells) mitosis.

Should excess growth hormone be secreted by young animals, perhaps due to a tumor in the pituitary, their growth would be excessive and would result in the production of a giant. Undersecretion of growth hormone in young animals results in stunted growth. Should a tumor arise in an adult animal after the actively growing cartilaginous areas of the long bones have disappeared, further growth in length is impossible. Instead, excessive secretion of growth hormone produces bone thickening in the face, fingers, and toes and can cause an overgrowth of other organs. Such a condition is known as acromegaly.

• PROBLEM 22:

Define the following terms: atom, isotope, ion. Could a single particle of matter be all three simultaneously?

Solution: An atom is the smallest particle of an element that can retain the chemical properties of that element. It is composed of a nucleus, which contains positively charged protons and neutral neutrons, around which negatively charged electrons revolve in orbits. For example, a helium atom contains 2 protons, 2 neutrons, and 2 electrons.

An ion is a positively or negatively charged atom or group of atoms. An ion which is negatively charged is called an anion, and a positively charged ion is called a cation.

Isotopes are alternate forms of the same chemical element. A chemical element is defined in terms of its atomic number, which is the number of protons in its nucleus. Isotopes of an element have the same number of protons as that element, but a different number of neutrons. Since atomic mass is determined by the number of protons plus neutrons, isotopes of the same element have varying atomic masses. For example, deuterium (^{2}H) is an isotope of hydrogen, and has one neutron and one proton in its nucleus. Hydrogen has only a proton and no neutrons in its nucleus.

A single particle can be an atom, an ion, and an isotope simultaneously. The simplest example is the hydrogen ion H^+. It is an atom which has lost one electron and thus developed a positive charge. Since it is charged, it is therefore an ion. A cation is a positively charged ion (i.e., H^+) and an anion is a negatively charged ion (i.e., Cl^-). If one compares its atomic number (1) with that of deuterium (1), it is seen that although they have different atomic masses, since their atomic numbers are the same, they must be isotopes of one another.

- **PROBLEM 23:**

What are the three laws of thermodynamics? Discuss their biological significance.

Solution: The first law of thermodynamics states that energy can be converted from one form into another, but it cannot be created or destroyed. In other words, the energy of a closed system is constant. Thus, the first law is simply a statement of the law of conservation of energy.

The second law of thermodynamics states that the total entropy (a measure of the disorder or randomness of a system) of the universe is increasing. This is characterized by a decrease in the free energy, which is the energy available to do work. Thus, any spontaneous change that occurs (chemical, physical, or biological) will tend to increase the entropy of the universe.

The third law of thermodynamics refers to a completely ordered system, particularly, a perfect crystal. It states that a perfect crystal at absolute zero (0 Kelvin) would have perfect order, and therefore its entropy would be zero.

These three laws affect the biological as well as the chemical and physical world. Living cells do their work by using the energy stored in chemical bonds. The first law of thermodynamics states that every chemical bond in a given molecule contains an amount of energy equal to the amount that was necessary to link the atoms together. Thus, living cells are both transducers that turn other forms of energy into chemical bond energy, and liberators that free this energy by utilizing the chemical bond energy to do work. Considering that a living organism is a storehouse of potential chemical energy due to the many millions of atoms bonded together in each cell, it might appear that the same energy could be passed continuously from organism to organism with no required extracellular energy source. However, deeper consideration shows this to be false. The second law of thermodynamics tells us that every energy transformation results in a reduction in the usable or free energy of the system. Consequently, there is a steady increase in the amount of energy that is unavailable to do work (an increase in entropy). In addition, energy is constantly being passed from living organisms to nonliving matter (e.g., when you write you expend energy to move the pencil, when it is cold out your body loses heat to warm the air, etc.). The system of living organisms thus cannot be a static energy system, and must be replenished by energy derived from the nonliving world.

The second law of thermodynamics is also helpful in explaining the loss of energy from the system at each successive trophic level in a food pyramid. In the following food pyramid, the energy at the producer level is greater than the energy at the consumer I level which is greater than the energy of the consumer II level.

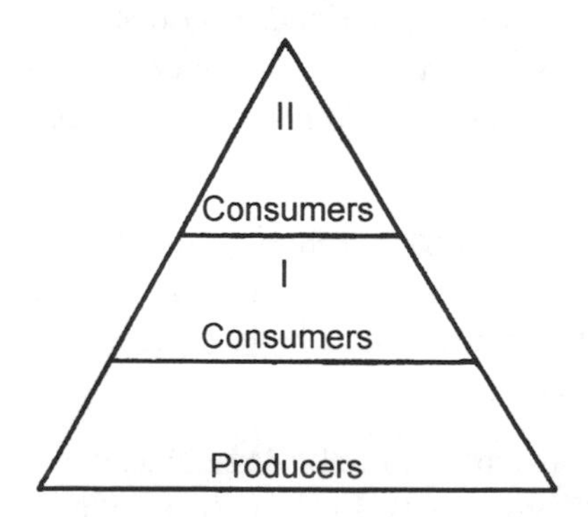

Every energy transformation between the members of the successive levels involves the loss of usable energy, and this loss serves to increase the entropy. Thus, this unavoidable loss causes the total amount of energy at each trophic level to be lower than at the preceding level.

• PROBLEM 24:

Why are the mitochondria referred to as the "powerhouse of the cell"?

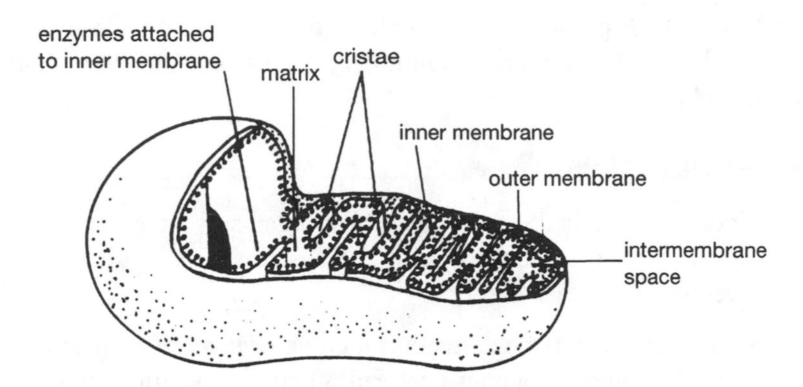

Diagram showing the internal structures of a mitochondrion through a cutaway view

<u>Solution</u>: Mitochondria are membrane-bound organelles concerned principally with the generation of energy to support the various forms of chemical and mechanical work carried out by the cell. Mitochondria are distributed throughout the cell, because all parts of it require energy. Mitochondria tend to be most numerous in regions of the cell that consume large amounts of energy and more abundant in cells that require a great deal of energy (for example, muscle and sperm cells).

Mitochondria are enclosed by two membranes. The outer one is a continuous delimiting membrane. The inner membrane is thrown into many folds, called cristae, that extend into the interior of the mitochondrion. Enclosed by the inner membrane is the ground substance termed the matrix (see accompanying diagram). Many enzymes involved in the Krebs cycle are found in the matrix. Enzymes involved in the generation of ATP by the oxidation of $NADH_2$, or, the electron transport reactions, are tightly bound to the inner mitochondrial membrane. The enzymes for the specific pathways are arranged in sequential orders so that the products of one reaction do not have to travel far before they are likely to encounter the enzymes catalyzing the next reaction. This promotes a highly efficient energy production.

The reactions that occur in the mitochondria are all related in that they result in the production of ATP (adenosine triphosphate), which is the common currency of energy conversion in the cell. Some ATP is produced by reactions that occur in the cytoplasm, but about 95 percent of all ATP produced in the cell is in the mitochondria. For this reason the mitochondria are commonly referred to as the powerhouse of the cell.

- **PROBLEM 25:**

Ordinarily, the bodies of multicellular organisms are organized on the basis of tissues, organs, and systems. Distinguish between these terms.

Solution: A tissue may be defined as a group or layer of similarly specialized cells which together perform a certain specific function or functions. Each kind of tissue is composed of cells which have a characteristic size, shape, and arrangement, and which are bound together by an intracellular substance allowing communication between adjacent cells. Some examples of tissues are epithelial tissue, which separates an organism from its external environments; muscle tissue, which contracts and relaxes; and nervous tissue, which is specialized to conduct information.

An organ is composed of various combinations of tissues, grouped together into a structural and functional unit. An organ can do a specific job. For example, the heart pumps blood throughout the circulatory system. By contrast, an organ can have several different functions. The human liver, for example, produces bile salts, breaks down red blood cells, and stores glucose as glycogen.

A group of organs interacting and cooperating as a functional complex in the life of an organism is termed an organ system. In the human and other vertebrates, the organ systems are as follows:

(1) the *circulatory system,* which is the internal transport system of animals;

(2) the *respiratory system,* which provides a means for gas exchange between the blood and the environment;

(3) the *digestive system,* which functions in procuring and processing nutrients;

(4) the *excretory system,* which eliminates the waste products of metabolism;

(5) the *integumentary system,* which covers and protects the entire body;

(6) the *skeletal system,* which supports the body and provides for body shape and locomotion;

(7) the *muscular system,* which functions in movement and locomotions;

(8) the *nervous system,* which is a control system essential in coordinating and integrating the activities of the other systems with themselves and with the external environment;

(9) the *endocrine system,* which serves as an additional coordinator of the body functions; and

(10) the *reproductive system,* which functions in the production of new individuals for the continuation of the species.

• **PROBLEM 26:**

Differentiate clearly between diffusion, dialysis, and osmosis.

Solution: Diffusion is the general term for the net movement of the particles of a substance from a region where the substance is at a high concentration to regions where the substance is at a low concentration. The particles are in constant random motion with their speed being directly related to their size and the temperature. When the movements of all the particles are considered jointly, there is a net movement away from the region of high concentration towards regions of low concentration. This results in the particles of a given substance distributing themselves with relatively uniform density or concentration within any available space. Diffusion tends to be faster in gases than in liquids, and much slower in solids.

The movement or diffusion of water or solvent molecules through a semipermeable membrane is called osmosis. The osmotic pressure of a solution is a measure of the tendency of the solvent to enter the solution by osmosis. The more concentrated a solution, the more water will tend to move into it, and the higher is its osmotic pressure.

The diffusion of a dissolved substance through a semipermeable membrane is termed dialysis. Dialysis is the movement of the solute, while osmosis is the movement of the solvent through a semipermeable membrane. Dialysis and osmosis are just two special forms of diffusion.

• PROBLEM 27:

Why is the phenomenon of diffusion important to movement of materials in living cells?

Solution: In a living cell, chemical reactions are constantly taking place to produce the energy or organic compounds needed to maintain life. The reacting materials of chemical reactions must be supplied continuously to the actively metabolizing cell, and the products distributed to other parts of the cell where they are needed or lower in concentration. This is extremely important because if the reactants are not supplied, the reaction ceases, and if the products are not distributed but instead accumulate near the site of reaction, Le Châtelier's principle of chemical reactions operates to drive the reversible reaction backward, diminishing the concentration of the products. Thus, in order to maintain a constant chemical reaction, the reactants must be continuously supplied and the products must move through the cell medium to other sites. Diffusion is how these processes occur.

When a certain chemical reaction is operating in the cell, some reacting substance will be used up. The concentration of this substance is necessarily lower in regions closer to the site of reaction than regions farther away from it. Under this condition, a concentration gradient is established. The concentration gradient causes the movement of molecules of this substance from a region of higher concentration to a region of lower concentration, or the reaction site. This movement is called diffusion. Thus, by diffusion, molecules tend to move to regions in the cell where they are being consumed. The products of the reaction travel away from the reaction site also by this process of diffusion. At the reaction site, the concentration of the products is highest, hence the products tend to move away from this region to ones where they are lower in concentration. The removal of products signals the reaction to keep on going. When the product concentration gets too high, the reaction is inhibited by a built-in feedback mechanism.

Thus, diffusion explains how movement of chemical substances occurs into or out of the cell and within the cell. For example, oxygen molecules are directed by a concentration gradient to enter the cell and move toward the mitochondria. This is because oxygen concentration is necessarily the lowest in the mitochondria where oxidation reactions continually consume oxygen. Carbon dioxide is produced when an acetyl unit is completely oxidized in the citric acid cycle. The CO_2 will then travel away from the mitochondria, where it is produced, to other parts of the cell, or out of the cell into the bloodstream where it is lower in concentration.

• PROBLEM 28:

What structural advantages enable mitochondria to be efficient metabolic organelles?

Solution: Mitochondria are called the "powerhouses" of the cell because they are the major sites of ATP production in the cell. They contain the enzymes involved in both the Krebs (citric acid) cycle and the electron transport system, which work together to furnish 34 of the 36 ATPs produced by the complete oxidation of glucose. Another 2 ATPs are produced by glycolysis, which occurs in the cytoplasm, making the total 36 ATPs produced for each oxidized molecule of glucose.

Pyruvate from glycolysis traverses both the outer and inner mitochondrial membranes easily because the membranes are totally permeable to pyruvate. Once inside the mitochondrion's matrix pyruvate is converted by the enzyme pyruvate dehydrogenase to acetyl coenzyme A. The coenzyme can then enter the Krebs cycle which occurs in the matrix where all the enzymes that catalyze the cycle's reactions are present (except succinic dehydrogenase which is located on the inner membrane). Conveniently bound to the surface of the cristae, the greatly folded inner membrane of the mitochondria, are the enzymes of the electron transport system. Since the cristae jut into the matrix, the distance needed to travel by $FADH_2$ and NADH, generated by the TCA cycle, to the enzymes of the electron transport system is reduced. Since the surface area of the inner membrane is so greatly enhanced by its cristae, the probabilities of NADH and $FADH_2$ encountering the electron transport system's enzymes which are concerned with ATP production become sharply increased. Also contributing to the efficiency of this organelle's ATP producing ability is the nature of the cristae's enzymes: they are actually multi-enzyme complexes which group together a number of sequentially acting enzymes responsible for electron transport and oxidative phosphorylation. These assemblies enhance the efficiency of respiration, since the product of one reaction is located near the enzyme of the subsequent reaction.

- **PROBLEM 29:**

Differentiate between the several types of heterotrophic nutrition and give an example of each.

Solution: The different types of heterotrophic nutrition are defined according to either the type of food source used or the methods employed by the organism in utilizing the food to obtain energy. Holozoic nutrition is the process employed by most animals. In this process, food that is ingested as a solid particle is digested and absorbed. Holozoic nutrition can be further classified as to the food source: herbivores, such as cows, obtain food from plants; carnivores, such as wolves, obtain nutrients from other animals; omnivores, like man, utilize both plants and animals for food.

Saprophytic nutrition is utilized by yeasts, fungi, and most bacteria. These organisms cannot ingest solid food; instead they must absorb organic material through the cell membrane. They live where there are decomposing bodies of animals or plants, or where masses of plant or animal by-products are found. A saprophyte obtains nutrients from nonliving organic matter. An example of a saprophyte is given by yeasts which produce ethanol. Utilizing grape sugar as their energy source, they ferment glucose to carbon dioxide and ethanol. These yeasts are used to produce wine.

In parasitic nutrition, organisms called parasites obtain nourishment from a living host organism. Most parasites absorb organic material and are unable to digest a solid particle. This is true also of saprophytes, however, saprophytes do not require a living host in order to supply them with nourishment. Parasites are found in many classes of the plant and animal kingdoms, and frequently are bacteria, fungi, or protozoa. All viruses are parasites, requiring the host not only for nutrition, but also for synthetic and reproductive machinery. Some parasites exist in the host causing little or no harm. Parasites causing damage to the host are well known to man, and are termed pathogenic. Examples of these are the tapeworm, which lives in the intestine and prevents the host from obtaining adequate nutrition from the food which is eaten, and the tubercle bacillus which causes tuberculosis. Certain organisms are saprophytic in their natural habitat, but are capable of living in a host organism and causing disease. *Clostridium tetani* is an example. In forests, these bacteria obtain nutrients from decaying plant and animal material. When the bacteria enter a wound in a human, the toxic substances they release cause the disease tetanus.

• PROBLEM 30:

You have decided to build up your body, so you have a dinner that is heavy in protein content, followed by a quart of milk. Trace the movement of the food from the mouth to the duodenum. What happens to the food in the stomach and small intestine?

Solution: When food is introduced into the mouth, it is mixed with saliva and masticated by the teeth into small pieces to facilitate swallowing and digestion (by increasing the surface area exposed to digestive enzymes). With the help of the tongue, the chewed food is rolled into a ball, or bolus, which is then thrust into the back of the mouth for swallowing.

As soon as the food enters the esophagus, it is moved down the tract by rhythmic muscular contractions known as peristaltic waves. Peristalsis is the alternate contraction and relaxation of the smooth muscles lining the digestive tract, and helps to move the bolus along. Movement of the bolus is also assisted in this region by salivary mucin.

As the food nears the stomach, the sphincter muscle controlling the opening of the stomach relaxes and allows the food to enter the stomach. In the stomach, it is acted upon by the gastric juice, which works optimally in an acidic medium. Gastric juice contains:

(i) *hydrochloric acid,* which acidifies the medium, in order for the proteolytic enzymes to work optimally;

(ii) *pepsin,* a proteolytic enzyme that cleaves long protein molecules into shorter fragments called peptides; and

(iii) *rennin,* an enzyme that solidifies casein, a milk protein, so that it can be retained in the stomach long enough for pepsin to act upon it.

After about four hours in the stomach, the semi-digested food, or chyme, is released into the duodenum through the pyloric sphincter. The duodenum receives pancreatic juice from the pancreas and bile from the gallbladder. The proteases trypsin and chymotrypsin, found in the pancreatic juice, further degrade peptides into smaller fragments.

The digestion of protein is then completed in the ileum, the last part of the small intestine. Here, the remaining peptide fragments are further digested to free amino acids by carboxypeptidases and aminopeptidases, which split off amino acids from the carboxyl and amino ends of the peptide chains, respectively. The free amino acids are then actively transported across the intestinal wall into the bloodstream.

• **PROBLEM 31:**

Describe the processes by which a root absorbs water and salts from the surrounding soil.

Solution: The movement of water from the soil into the root can be explained by purely physical principles. The water available to plants is present as a thin film loosely held to the soil particles and is called capillary water. The capillary water usually contains some dissolved inorganic salts and perhaps some organic compounds, but the concentration of these solutes in capillary water is lower than that inside the cells of the root. The cell sap in the root hair of the epidermal cells has a fairly high concentration of glucose and other organic compounds. Since the plasma membrane of this cell is semipermeable (it is permeable to water but not to glucose and other organic molecules), water tends to diffuse through the membrane from a region of higher concentration (the capillary water of the soil) to a region of lower concentration (the cell sap of the root hair). This movement of water is controlled by a process called osmosis.

As an epidermal (root hair) cell takes in water, its cell sap now has a lower solute concentration than that of the adjacent cortical cell. By the same process of osmosis, water passes from the root hair to the cortical cell. Because of the osmotic gradient, water will continue to diffuse inward toward the center of the root. In this way, water finally reaches the xylem and from there, water is transported upwards to the stem and leaves by a combination of root pressure and transpiration pull.

• **PROBLEM 32:**

What is meant by translocation? What theories have been advanced to explain translocation in plants? Discuss the value and weakness of each.

Solution: Translocation is the movement of nutrients from the leaves where they are synthesized to other parts of the plant body where they are needed for a wide variety of metabolic activities. Translocation takes place through the sieve tubes of the phloem.

Experimental data indicates that the high rate at which translocation occurs cannot be attributed to simple diffusion. Three theories have been offered to explain the mechanism of translocation. Each of them has its own value and weakness, but the first one presented (pressure-flow theory) holds the most support.

The pressure-flow theory, which had been widely accepted in the past, proposes that the nutrient sap moves as a result of differences in turgor pressure.

Sieve tubes of the phloem in the leaf contain a high concentration of sugars, which results in high osmotic pressure into the cells. This osmotic pressure causes an influx of water, and build up of turgor pressure against the walls of the sieve tube cells.

In the roots however, sugars are constantly being removed. Here, a lower concentration of solute is present resulting in a lowering of the osmotic pressure. There is consequently a lower turgor pressure exerted against the sieve tube walls of the root as compared to the leaf. This difference in turgor pressure along the different regions at the phloem is believed to bring about the mass flow of nutrients from a region of high turgor pressure such as the leaves (where photosynthesis produces osmotically-active substances like glucose) to regions of lower turgor pressure such as the stem and roots. The fluid containing the nutrients is pushed by adjacent cells, along the gradient of decreasing turgor, from the leaves to the roots. This theory predicts that the sap should be under pressure as it moves down the phloem, and this is experimentally verified. But there are problems with this hypothesis. Under some conditions, sugar is clearly transported from cells of lesser turgor to cells of greater turgor. In addition, this theory fails to explain how two substances can flow along the phloem in different directions at the same time, a situation observed to occur by some investigators.

Another theory proposes that cyclosis, the streaming movement evident in many plant cells, is responsible for translocation. According to this theory, materials pass into one end of a sieve tube through the sieve plate and are picked up by the cytoplasm which streams up one side of the cell and down the other. At the other end of the sieve tube, the material passes across the sieve plate to the next adjacent sieve tube by diffusion or active transport. This theory is able to account for the simultaneous flowing of nutrient saps in different directions. However, it is attacked by some investigators on the basis that cyclosis has not been observed in mature sieve cells.

A third theory proposes that adjacent sieve-tube cells are connected by cytoplasmic tubules, in which sugars and other substances pass from cell to cell. The movement of these substances is powered by the ATP from the mitochondria-like particles that are believed to lie within these connecting tubules. This theory is, however, weak since it is supported by few experimental findings and much of its content is based on speculation.